Second Edition

INDUSTRIAL FIRE PROTECTION HANDBOOK

Second Edition

INDUSTRIAL FIRE PROTECTION HANDBOOK

R. Craig Schroll

CRC PRESS

Boca Raton   London   New York   Washington, D.C.

Library of Congress Cataloging-in-Publication Data

Schroll, R. Craig.
    Industrial fire protection handbook / R. Craig Schroll.—2nd ed.
        p. ; cm.
    Includes bibliographical references and index.
    ISBN 1-58716-058-7 (alk. paper)
    1. Industrial buildings—Fires and fire prevention. I. Title.

TH9445.M4 S37 2002
628.9′22—dc21   2002018808
                                                                        CIP

Visit the CRC Press Web site at www.crcpress.com

© 2002 by CRC Press LLC

No claim to original U.S. Government works
International Standard Book Number 1-58716-058-7
Library of Congress Card Number 2002018808
Printed in the United States of America   2 3 4 5 6 7 8 9 0
Printed on acid-free paper

Dedication

To my wife Penny for her support of my work in spite of the time commitment involved. To my parents Bert and Roy, whose support over the years has played a major role in shaping me.

Preface

This book is designed to provide practical guidance that may be implemented to improve fire prevention and protection within a work environment. The emphasis is on techniques and methods that can be applied to improve actual performance. The book is intended for both the student who has yet to begin practicing in the field and for the practitioner who has chosen to be a student to expand his or her knowledge of industrial fire protection. The primary focus of this book is fire protection in an industrial setting; however, much of the material applies to fire protection issues in any type of occupancy.

Fire loss prevention and control are fundamentally about developing systems and practices within a facility to increase the opportunities to avoid fires, limit the development and spread of fires, and allow for the rapid and effective control of fires.

While codes and standards are occasionally referenced, this book is not concerned with the detailed requirements of these documents. There are engineering aspects included herein, but the goal is not to make each reader an engineer.

If your goal is to focus exclusively on fire protection in your career, this text will provide a good introduction to the body of knowledge involved. You should follow this with additional detail on each of the specific areas. If you are a general practitioner in the field of safety, this text will provide the fundamental information you need to effectively handle your fire related responsibilities.

Additional illustrations are used throughout this second edition to improve understanding of many of the issues involved with effective fire loss prevention and control. The introductory information and examples in Chapter 1 have been updated from the previous edition. Chapter 2 provides significant information on new extinguishing agents including wet chemical and clean agents designed to replace halon. The general loss control program coverage in Chapter 3 has been significantly expanded. Life safety issues in Chapter 4 have been covered in more detail, and several additional examples have been included. The hazard control discussion in Chapter 5 has been expanded to include more in-depth coverage of the issues dealt with in the first edition as well as additional issues. Chapter 6's coverage of installed fire protection systems has been expanded. Additional details on the systems covered in the first edition are included and discussion of a broader array of systems has been added. Chapter 7 on portable fire extinguishers has been improved, and new material about the latest extinguisher types has been added. Additional examples covering selection, placement, and maintenance have been included. The coverage of emergency planning in Chapter 8 has been improved, and the emergency team and fire brigade coverage in Chapter 9 has been enhanced with new information. Chapter 10 on coping with a fire has been updated and expanded. Finally, the appendix materials have been changed completely to reflect the widespread availability of information on the Internet.

Acknowledgments

Many people affect an individual in the course of a career. I would like to take this opportunity to acknowledge some of the most noteworthy contributors to my professional development. I owe my beginnings in the fire service to Paul Wise, retired Chief of the Liberty Fire Company in New Holland, Pennsylvania. His efforts during my first exposure to fire protection helped me select this as my career. I owe a debt of gratitude to Chief Walter J. McNeil, who was my first supervisor as a paid fire protection specialist in the U.S. Air Force. He suffered occasionally at the hands of an often overzealous rookie, but he took it all in stride. He was always helpful, encouraging, and provided an excellent example of what a professional was supposed to be. I extend my thanks to the many colleagues I have been fortunate enough to work with while participating in various professional associations. They have given me an opportunity to contribute and have added much to my experience over the years. Last, but far from least, I thank all of my students and clients for the opportunities they have given me and the source of learning and growth they have provided that has made this book possible.

If you have suggestions or comments regarding this book, please e-mail them to the author at Craig@FIRECON.com or send them to FIRECON, P.O. Box 231, East Earl, PA 17519, USA.

List of Figures

Note: All figures not otherwise attributed are ©Firecon and printed with permission from Firecon.

Table of Contents

Chapter 4   Life Safety

Chapter 5   Hazard Control

Chapter 6   Installed Fire Protection

Chapter 7   Portable Fire Extinguishers

Chapter 10 Coping with Fire

Appendix A: Annotated Bibliography 229

Appendix B: Resource List. 231

Index 239

1 Nature of the Problem

CHAPTER OBJECTIVES

You will be able to identify and explain:

- Scope of the fire loss problem
- Specific nature of fire loss potential
- Trends in fire loss
- Factors which affect fire loss
- Areas where loss control personnel can have a positive impact on losses

SCOPE OF THE FIRE LOSS PROBLEM

Understanding the nature and scope of the fire loss problem is necessary to provide a basis for reducing fire losses. Losses offer valuable lessons because they provide the information needed to prevent or reduce the impact of similar losses in the future. While this chapter discusses the past, it more importantly presents the concept of learning from the past and applying this knowledge to future prevention.

PAST LOSSES

Losses from fire have escalated over the last century. In 1910, the per capita fire loss was $2.32; by 1999, this figure had risen to $36.80. The total cost of fire loss is even higher. Total fire costs include not only direct fire damage but also the costs of preventing and controlling fires. While the cost of fire losses has continued to rise, the number of deaths from fire has, fortunately, declined slightly.

In 1999, there were 523,000 structural fires in the United States. The estimated direct property loss was $10.3 billion. An estimated 3570 civilians were killed in fires and 21,875 were injured. The vast majority of these injuries and fatalities occurred in residential fires.

LOSS POTENTIAL

Loss potential refers to the magnitude of the loss or losses which are reasonably possible and which are likely to occur at some point in time. Losses and their costs are typically divided into human, direct, and indirect components.

Deaths of and injuries to personnel are a portion of human loss from an uncontrolled fire. Human loss also concerns nonmeasurable items such as pain, suffering, and grief. The costs also include measurable items such as hospital expenses, disability compensation, and other benefits. Direct losses, such as burnt equipment, material, and buildings, as well as smoke and water damage, are an immediate result of fire. These losses can be seen and counted relatively easily. Indirect losses include the expenses of lost production time, business interruption, reproduction of records, and many more. Indirect losses can, to an extent, be measured but generally cannot be seen. These losses are a direct result of the fire but are often overlooked when calculating the true cost of fire losses.

The resources expended to deal with fire losses should also be considered part of the cost of fire. The cost of installed fire protection, fire departments, and insurance premiums, to name a few, are all resources devoted to dealing with uncontrolled fires.

Most organizations have a greater loss potential than the management of the organization realizes. Individuals tend to believe that things like a major fire loss will not happen to them personally. The statistics presented here, however, vary only slightly from previous years and offer a fair projection of next year's numbers. These losses occur to someone. It is safe to assume that, without effective action, someone could just as easily be you as anyone else. This is an important point. To effectively work for the prevention and control of losses, it is necessary to believe that a loss is possible. One thing that often precedes a significant loss is the complacency brought on by not having yet experienced a loss.

CASE HISTORIES

The following case histories are actual incidents and events used to illustrate more fully the nature of the fire loss control challenge. The purpose of these cases is to demonstrate how losses occur and to introduce the concept of learning from the experiences of others. The cases presented here are designed to provide educational illustrations; they are not intended to place blame or assign responsibility for the outcomes of the incidents. Hopefully, over time, actions that have been successful are repeated, and ones that have failed are avoided. Everyone should take the time to learn from the experiences of others. This experience provides a valuable source of information that is considerably less costly than personal experience.

CASE ONE[1]

Shortly after noon on April 16, 1984, welding operations were being performed on equipment in a plywood manufacturing plant. In the area where the maintenance work was taking place, deposits of oil, pitch, and wood dust were present on equipment, structural members, and catwalks. The welding operations ignited a fire in these deposits which spread rapidly. The fire department arrived approximately five minutes after the fire ignited and found the plant heavily involved in fire. Within another five minutes, the structure was fully involved in fire. Ten minutes later, the roof began to collapse.

The fire caused an estimated direct loss of $32.5 million. Three major factors were identified as contributing to the loss: (1) combustible deposits of oil, pitch, and wood dust, (2) lack of adequate fire prevention during welding, and (3) obstruction of piping in the installed fire protection system.

The plant was completed in 1970 and included a boiler house, cooling shed, planing mill, stud mill, and the main plywood production building. The main plywood manufacturing building was constructed of wood and contained over 200,000 square feet (18,580 square meters) of undivided floor area.

The plant building was protected by 12 dry-pipe sprinkler systems and two dry-pipe standpipe systems which supplied 30 hose stations. The water supply to these systems was limited to a municipal water connection. A 10-inch (25.4 cm) main fire protection loop surrounded the plant, and this loop was fed by an 8-inch (20.3 cm) and 12-inch (30.5 cm) city water main. Each system connection to the fire loop was through a post indicator valve. These valves were locked open and inspected weekly. The plant fire brigade had received training in the use of standpipe hoses and had controlled several previous fires prior to the arrival of the fire department.

The welding was to be performed at a point approximately fifteen feet above floor level on a piece of production equipment. The area surrounding the welding work was wetted down with a 0.75-inch (1.9 cm) garden hose at floor level and a fire watch was posted. Combustible materials at the level of welding and above were not wetted down. Shortly after welding operations were started, the welder sensed heat coming from above him. The employee discovered that a fire had started to burn on the catwalk and equipment frame above. The fire watch then tried to control the fire with the 0.75-inch (1.9 cm) hose. This hose did not provide sufficient reach or water volume, however, and was ineffective. The fire spread rapidly to the top of the equipment and into a draft hood above the equipment. An employee in a breakroom noticed the fire, sounded the alarm over the public address system, and called the public fire department. This employee attempted to use a 1.5-inch (3.8 cm) standpipe hose line, but when the line only spurted water, he assumed the line was inoperative and went to get a 2.5-inch (6.4 cm) hand-line from an outside hose cabinet. When he returned, the fire was beginning to break out of the draft hood and appeared too large to control. This handline was not used on the fire. None of the employees present during this period noticed any evidence of sprinkler water flow.

Approximately 20 minutes into the incident, the roof collapsed, which ruptured the sprinkler system piping and effectively eliminated any potential control of the fire. Fire department operations were primarily confined to defensive measures to prevent the spread of the fire to adjacent areas.

Although the plant had a hot work policy and established procedures for welding operations, these were not effective. The area surrounding the welding site was not properly secured prior to welding. Ideally, combustible materials near the work area in all directions must be considered. Fire watch personnel should be the first to discover a fire. In this case, the welder was the first to notice the fire, indicating that the fire watch person was not properly positioned to see all essential areas or was not as attentive as he should have been. More effective fire control measures should have been readily available. The 0.75-inch (1.9 cm) hose that was used did not provide the reach or necessary volume of water to handle a fire in this area. The accumulations of oil, pitch, and dust should have made the need for more adequate fire control capability evident.

The alert and evacuation operations were initiated promptly and carried out effectively. Calling the fire department immediately was also appropriate. These actions were handled well by the employees, and no one was injured as a result of this fire.

The operation of the 1.5-inch (3.8 cm) standpipe hose may have been effective because in dry-pipe systems there is a delay in obtaining full water flow at the nozzle while the air exhausts from the system. This delay could have been the only cause of the spurting observed by the employee. The hose line may have been sufficient to confine the fire to the area of the draft hood had it been operated. In post-fire investigations, deposits were found in the sprinkler and standpipe system piping. Internal deposits in fire protection system piping can significantly reduce water flow and may have been partially to blame for the minimal water flow. These deposits were also responsible for inadequate performance of the sprinkler system. Fire protection system maintenance is critical. Effective maintenance and inspection procedures should have revealed the blockage of system piping.

Fire loss experience documented by Factory Mutual over the past fifteen years indicates that fires in structures with wood roofs, which have initiated the operation of more than 20 sprinkler heads, have been successfully contained or controlled by properly operating sprinkler systems.

CASE TWO[2]

On May 27, 1987, a small accidental spill of flammable liquid was ignited by sparks from a forklift. The resulting fire consumed 1.5 million gallons (5.678 million liters) of paint and automotive flammable liquids and destroyed the warehouse in which they were stored. The direct loss from this fire was $49 million.

Construction of the facility was completed in 1977. The main structure contained the warehouse, an office area was attached, and a covered drum storage area was adjacent to the building. A fire wall divided the warehouse into two areas: the east side had approximately 98,600 square feet (9160 square meters) of useable area and the west side had about 82,000 square feet (7618 square meters). The fire wall should have provided 4-hour fire resistance based on the construction. Four openings were provided in the fire wall for vehicle traffic; each of these openings was protected with a 3-hour sliding fire door on both sides of the opening. There were two personal doors through the fire wall which were also protected by 3-hour doors.

The facility was protected by eight wet-pipe sprinkler systems and one deluge system. A standpipe system and hose stations were provided throughout the warehouse. Installed systems were supplied with water from a 16-inch (40.6 cm) municipal water main. A 2500-gallon-per-minute (GPM) (9463 liters/minute) fire pump was located on-site in a separate pump house and supplied a 10-inch (25.4 cm) fire protection loop. Sprinklers were also provided in storage racks. Standpipe hose stations were present throughout the facility. Manual alarm pull stations were located next to exits. The sprinkler, standpipe, and manual alarm systems were monitored by a central station. Fire extinguishers were provided in all areas of the warehouse.

The company maintained an employee emergency action team of 18 people per shift. These individuals were trained in fire control with extinguishers and standpipe hoses, spill control, and other basic tasks.

A lift truck operator was in the process of placing two pallets of a lacquer-flattening agent on top of two pallets already in the stack. The load shifted and some of the material fell off the forklift. Several cartons struck the forklift operator's cage. Containers opened and spilled flammable liquid over the forklift and operator. Other materials hit the floor and opened, creating a spill approximately 12 feet (3.6 meters) in diameter.

Two employees in the area heard the accident. One reported it, and the other went to get the spill control cart. The emergency team was activated and responded. The lift operator attempted to stabilize the remaining load that was still elevated. During this process, sparks from the lift truck ignited the spilled flammable liquid.

Two emergency team members were preparing a 1.5-inch (3.8 cm) hose line when they saw the lift operator engulfed in flames. They rescued the operator and took him out of the building. Another employee attacked the fire with a fire extinguisher but was unable to extinguish it. A supervisor activated the building alarm. Emergency team members realized that the fire was too severe for their training and equipment and evacuated the building.

Municipal firefighters arrived approximately 9 minutes after ignition. The fire was already coming through the roof of the structure. The entire warehouse was fully engulfed in fire within 30 minutes after ignition.

One or both of the fire doors at each fire wall opening operated properly during the fire. The fire possibly spread past the fire wall due to flammable liquid running under the doors. Fire wall openings were not provided with curbs or drains. Although the sprinklers and automatic roof vents operated during the fire, they were not sufficient to control it.

The initial planning for this facility recognized the hazardous nature of the materials to be stored, and fire protection systems and procedures were adequate at first. During years of use, the types of containers and storage arrangement changed, but the fire protection systems were not evaluated and updated to compensate for these changes. The additional fuel load overwhelmed the fire protection systems.

CASE THREE[3]

The morning work shift of employees at a poultry processing plant had just begun when a fire occurred at approximately 8:15 a.m. on September 3, 1991. The rapid spread of heavy smoke throughout the structure ultimately resulted in 25 fatalities and 54 people being injured in varying degrees.

Imperial Foods occupied a one-story brick and metal structure that had been used over the years for various food product operations. The total area was approximately 30,000 square feet (2787 square meters).

Operations at this plant did not include the slaughter of poultry. Rather, poultry parts were shipped to the plant which then prepared and cooked the chicken.

The plant employed approximately 200 people, with a normal shift having around 90 employees. Preparation of the poultry products included trimming, marinating, cutting, and mixing. The prepared meat would then be cooked, quick-frozen, packed, and prepared for shipping.

Poultry products that had already gone through the various marinating and mixing procedures were taken by conveyor to a cooking vat in the processing room which contained soybean oil. The oil was maintained by a thermostat control at a constant temperature of 375°F (190.6°C) plus or minus 15°F (9.4°C).

A maintenance worker who survived the fire indicated that the hydraulic line that drove the conveyor had developed a leak. The hydraulic line was turned off and drained of fluid. Then the maintenance worker disconnected the leaking line and replaced it with a factory prepared line.

The factory prepared line, however, was found to be too long and would have dragged on the floor, possibly causing people working in the area to trip. So the maintenance worker reportedly asked for and gained permission to cut the factory prepared hydraulic line to an appropriate length, replaced the end connector with his own connector, and put the line back in place. This line has been described as a 0.75-inch (1.9 cm) flex line rated to carry 3000 psi (20,684 kilopascals). Information from plant personnel indicated normal pressure was kept at approximately 800 psi (5516 kilopascals) but would at times fluctuate as high as 1200 to 1500 psi (8274 to 10,342 kilopascals).

The hydraulic line was brought back to operating pressure. Shortly afterward, it separated at the repaired connector point. The connector was some 4 to 6 feet (1.2 to 1.8 meters) above floor level with hydraulic fluid being expelled at a pressure of 800 to 1500 psi (5516 to 10,342 kilopascals). It began to splatter off the concrete floor. Droplets were bouncing back onto the gas heating plumbs for the cooking vat, which turned them into vapor. The vapors were then going directly into the flame. The vapors had a much lower flashpoint than the liquid hydraulic fluid and, therefore, rapidly ignited.

The pressurization of the hydraulic fluid combined with the heat was causing an atomizing of the fuel which, in all probability, caused an immediate fireball in and around the failed hydraulic line and the heating plumbs. The ignition of the fuel caused an immediate and very rapid spreading of heavy black smoke throughout the building. Seven workers were trapped between the area of origin and any escapable routes.

Measurement of the system during the investigation after the fire indicated 50 to 55 gallons (189 to 208 liters) of hydraulic fluid fueled the fire before electrical failure shut the system down. In addition to the hydraulic fluid, the fire reached a natural gas regulator that, in turn, failed and caused an induction of natural gas to the fire increasing the intensity and buildup of toxic gases.

Witness reports indicated much of the plant was enveloped in flames in less than two minutes. Workers throughout the plant found their visibility eliminated and oxygen quickly consumed. Hydrocarbon-charged smoke, particularly as heavy as this, is extremely debilitating to the human body and can disable a person with one or two breaths. This was confirmed as autopsies conducted on all of the fatalities found that virtually all died of smoke inhalation as opposed to direct flame injury.

Survivors indicated that there was no real organization in the plant's evacuation, and this was confirmed by the locations of the bodies. Several employees in the central part of the structure moved to the trash compactor/loading dock area near the southeast corner of the building. It was there they found one of the personnel doors to the outside locked. A trailer was backed into the loading dock cutting off all potential to exit through this area. One woman became trapped between the compactor seal and the building wall while trying to squeeze through an opening. A number of remaining people in this area went into a large cooler adjacent to the loading dock but failed to pull the sealed door shut, thus allowing smoke infiltration into the cooler. The cooler had the largest single fatality count area with 12 deceased people being removed from this room along with five injured people.

The second largest concentration of fatalities was seven people trapped in the processing room between the fire and any escape route. Three additional bodies were found in the trim room area, one of whom was a route salesman who had been filling food machines in the break room. The exterior personnel door in the break room was the other door locked from the outside.

Several factors contributed to the seriousness of this fire and the accompanying large loss of life. A few of the most significant are mentioned here. Appropriate maintenance of equipment such as the hydraulic line is critical to fire prevention. Installed fire protection and other engineering controls could have minimized the impact of the fire. The locked fire exits and building layout prevented effective escape for a large number of employees.

CASE FOUR[4]

During a transfer of flammable liquid from one container to another, an individual spilled some on himself. In the course of going to the locker room to clean up, he passed by the rear of an operating powered industrial truck. The engine of the truck provided an ignition source for the vapors being released off the person's clothing and he received third-degree burns over 50% of his body.

FACTORS THAT AFFECT FIRE LOSS

Many factors are involved in and have a direct effect on any specific fire loss. However, several common denominators exist in most incidents. People are the first challenge to fire loss control efforts, and Chapter 3 will discuss their impact in more detail. For now, it is important to remember that people are a major factor in almost all losses and that they have a significant impact on all three of the factors discussed below. The factors influencing the severity of a fire loss can be divided into three basic categories. Many individual items fall within each of these three areas, but each item within a group has a similar effect on our fire losses.

The first area consists of factors that influence the start of a fire. All of the items in this category have a direct or indirect impact on fire ignition. To have a fire, an ignition source must be available in the presence of some type of fuel. There must be an act, omission, or system failure that juxtaposes these two items. Prevention efforts in this area are designed to avoid the ignition of a fire. Activities include a smoking policy, a hot work permit system, and equipment maintenance procedures.

The second major group of factors are those which contribute to the growth and spread of a fire. For continued fire growth and spread to occur, there must be sustaining quantities of fuel and avenues of fire spread. Items such as housekeeping and building construction have a major impact in this area.

Factors which assist with the control or extinguishment of fires are the third major group. These are the human or mechanical means of intervention that control the fire or limit losses. This area concerns employee training, installed fire protection systems, and hazard separation.

Most fire loss control efforts fall into one or more of these three categories. It will be helpful to use these categories when thinking about fire loss control options and alternatives.

CAN LOSS PREVENTION AND CONTROL HAVE AN IMPACT?

All three categories of factors which influence fire losses have aspects that can be controlled. The keys to our efforts as individuals responsible for loss control are the same as the keys to success in most endeavors: hard work and constant effort. If we intend to be effective in preventing, planning for, and controlling fire losses, we have to commit ourselves to that goal and expend the effort necessary to be successful. It is not always easy, but it is always rewarding.

Loss prevention is one of the few professions which measures its success by a failure rate, and, frequently, it is not possible to accurately determine how successful our efforts have been. There is a trend toward developing more effective and positive ways of measuring loss prevention performance, but for now we still measure loss prevention performance primarily with the yardstick of losses that have occurred.

REFERENCES

1. Timoney, T., Texas plywood manufacturing plant destroyed, *Fire J.*, 79(2), 53, 1985.
2. Isner, M., $49 million loss in Sherwin-Williams warehouse fire, *Fire J.*, 82(2), 65, 1988.
3. Technical Report, United States Fire Administration.
4. WorkSafe Western Australia, Web site.

2 Fire Behavior

CHAPTER OBJECTIVES

You will be able to identify and explain:

- What a fire is
- Necessary elements for a fire to occur
- Characteristics and types of fuel classes
- Characteristics and types of ignition sources
- Measures of flammability
- Characteristics of flammable liquids
- How heat is transferred during a fire
- Products of combustion
- Extinguishing agents and methods

You will be able to:

- Predict certain characteristics of a potential fire situation based on known factors
- Select the appropriate extinguishing agent and method given a specific fire situation

An understanding of fire behavior is an essential foundation to effectively prevent, plan for, and control fire losses. The need for a working knowledge of this important area is similar to a sports team watching films of another team they will be playing or a military tactician who studies the tactics used by his opponent. To effectively prevent a loss, we must understand what causes it. To effectively plan for losses, we must know how they occur, and, more importantly, why they occur. To control losses, we must have a grasp of the strengths and weaknesses of the enemy, in this case, fire.

Fire is a natural phenomenon and is controlled by certain natural laws. Since fire cannot act contrary to these laws, an understanding of the natural laws that govern fire behavior is a major advantage.

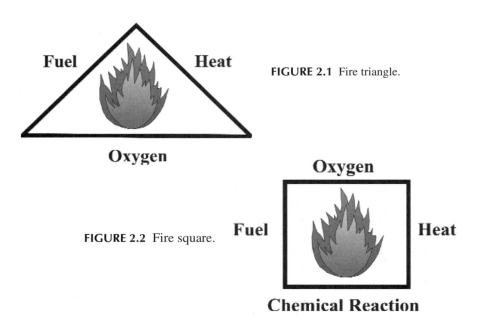

FIGURE 2.1 Fire triangle.

FIGURE 2.2 Fire square.

WHAT IS FIRE?

Fire is rapid, self-sustaining oxidation accompanied by the evolution of varying intensities of heat and light. This definition indicates that fire is a chemical process of decomposition in which the rapid oxidation of a fuel produces heat and light. This process makes fire the mid-range reaction based on the speed at which the two other common forms of oxidation occur. Rust, or corrosion, is an example of the slower form, and explosion is an example of the more rapid form.

ELEMENTS OF FIRE

Three basic elements must be present for a fire to occur: fuel, heat, and oxygen. These three components make up the fire triangle (Figure 2.1), and proper combination of these three items invariably results in a fire. The chemical chain reaction between the fuel, heat, and oxygen represents the fourth component of the fire equation. We will refer to this as the fire square (Figure 2.2). Anytime something burns, these four components are present. Preventing the combination of these elements will prevent a fire. If one of the elements is removed from the fire situation, the fire will be extinguished.

The elements are not fixed quantities, each has an impact on the others. A preheated fuel does not require as strong an ignition source as it would if it were not preheated. For example, if a gasoline spill on pavement were to occur when the temperature was close to zero (17.7°C), the gasoline would be less likely to ignite than if the same spill occurred on a 90°F (32°C) day. In an oxygen-enriched atmosphere, a fuel will become easier to ignite. These changes occur to all four of the elements involved in the process, and a change in one will have an impact on the others.

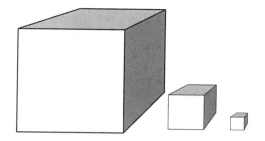

FIGURE 2.3 Surface-to-mass ratio.

The oxygen element of the fire triangle can be viewed more scientifically as an oxidizer. Some chemicals behave very much like oxygen. Chlorine, for example, may contribute to a fire in other materials because it is an oxidizer. Some materials, such as ammonium nitrate, contain enough oxygen within their chemical structure that external oxygen is not needed for a fire to occur.

The physical state of the fuel is also important. A solid wooden board is more difficult to ignite than wood shavings due to the mass to surface area ratio. If the mass is large and the surface area small, as with the solid board, the heat of an ignition source is more easily dissipated through the material. If the mass is small and the surface area large, as with wood shavings, the heat cannot be dissipated as quickly, and ignition occurs more easily.

For example, in Figure 2.3, a block of wood 3 by 3-inch square (7.6 by 7.6 cm) has a surface of 54 square inches (348 square cm). If this block of wood were cut into 1-inch (2.54 cm) cubes, the mass remains the same but now the surface area is 162 square inches (1045 square cm). If each one inch cube was further cut into 0.33 by 0.33-inch (0.84 cm) cubes, the mass still remains the same but now the surface area is 2187 square inches (14,110 square cm).

Dust is an example of reducing mass relative to the surface area. Given the proper conditions, many dusts may explode. Grain and coal dust are two common varieties of dust that can explode.

In a fire, it is the vapors that are actually burning, so the closer the fuel is to the vapor state, the easier it is to ignite. Liquids ignite more readily than solids, gases more easily than liquids. These physical states affect many aspects of our loss control efforts. It is important to remember that any change in the physical state of a material may also change the characteristics and behavior of that material under certain conditions.

CLASSES OF FIRE

Fire is divided into five classes based primarily on the fuel that is burning. This classification system helps us assess hazards and determine the most effective type of extinguishing agent. It is also used for classifying, rating, and testing fire extinguishers. The older style class symbols are shown in Figure 2.4 and the newer symbols in Figure 2.5. A class K symbol is shown in Figure 2.6.

Class A fires involve ordinary combustible materials: wood, paper, and cloth. Class A fires are usually relatively slow in their initial development and growth, and

FIGURE 2.4 Old-style class symbols.

FIGURE 2.5 New class
pictographs.

FIGURE 2.6 Class K symbol.

because these materials are solids, they are somewhat easier to contain. Class A fires leave an ash after the material has been consumed.

Class B fires involve flammable and combustible liquids and gases such as gasoline, fuel oil, and propane. These fires usually develop and grow very rapidly. Class B materials are fluid in nature, which allows them to flow and move. This makes dealing with them somewhat more difficult than Class A materials. These materials are common in many settings. These fires typically do not leave an ash.

Class C fires involve energized electrical equipment such as motors, appliances, and machinery. This is the only classification of the five that is not directly related to the type of fuel. The fact that a live electrical circuit is involved is the determining factor. The burning materials may actually fall into one or more of the four other classes. If the electric power is disconnected, the fire is no longer considered class C. Whether the device being considered is turned on or not is unimportant in this classification. Power to the device makes it Class C even if the device is turned off.

Class D fires involve combustible metals such as magnesium, titanium, and zirconium. These materials are usually difficult to ignite but create intense fires once started. Class D fires are very difficult to extinguish, but, fortunately, they are relatively uncommon in most industries.

Class K fires involve cooking oils. This is the newest of the fire classes.

IGNITION SOURCES

Ignition sources can be grouped into four categories based on the type of source involved in the heat-producing action. These categories are: mechanical (friction, compression); electrical (resistance, arcing, static, lightning); chemical (combustion, decomposition, spontaneous heating, solution); and nuclear.

Mechanical heating occurs in two forms: friction and compression. Friction between two surfaces that are in contact with at least one surface in motion creates a heat build-up which can lead to the ignition of combustible materials. A simple example is a belt slipping on a pulley. The action of the belt moving against the pulley generates heat, and the belt, being a combustible material, provides the fuel. Friction can also result in ignition if the friction produces sparks. For example, sparks from a grinding operation may ignite combustibles in the area. Compression is not a common source of ignition. The gas laws of chemistry state that anytime the pressure on a gas is increased, the temperature also increases. It is possible that during the compression of a gas, sufficient heat could be generated to cause ignition of combustible materials. This principle is one of the reasons why a diesel engine works without spark plugs.

Electrical heating occurs in several ways, but we are going to limit our discussion to the more common electrical ignition sources: resistance, arcing, static electricity, and lightning. Resistance in an electric circuit is very similar to friction and generates heat. As an electric current passes through a conductor, a certain amount of the current is needed to overcome the friction between atoms as the electricity passes from one to the other. This causes a resistance to the passage of the current which, in turn, produces heat. How much heat is produced varies depending on a number of factors such as whether the conductor is insulated or not, how big it is, what the conductor is made from, and how quickly it dissipates the heat. An example of intentional resistance heating is the burner on an electric stove. We run into problems when these principles begin to work on a conductor not intended for heating. If a conductor is caused to carry more current than it was designed for, heating will occur that may cause a fire.

Arcing occurs anytime a circuit carrying electric current is interrupted. The arc is created by the current's tendency to maintain the established flow. The magnitude of the problem created by the arc depends on the current being carried and the method of interruption of the circuit. Disconnecting a switch creates a small arc that is usually not a problem. A conductor placed in close proximity to a grounded object is likely to create an arc that would be a problem if combustibles were nearby.

Static electricity involves the formation of an electric charge on the surface of each of two materials, one positive and the other negative. Given the proper circumstances, these two charges will cause an arc between the surfaces of the two materials. If the two materials happen to be gasoline and the inside surface of a container into which it is being dispensed, a fire may result. Any time dissimilar materials are moved in close proximity, static may be an issue. Flammable liquids being pumped through a pipe or hose, plastic pellets being blown through a duct, and plastic film being drawn through a printing press are all examples of areas where static may be a concern.

The only fire cause that can truly be described as an "act of God" is lightning. Lightning involves the build-up of charges between clouds or between clouds and the earth's surface. When these charges reach a point at which sufficient energy is present, they discharge and create lightning. This produces a tremendous amount of heat.

Chemical heating comes about in four ways: combustion, decomposition, spontaneous heating, and solution. Combustion is a relatively simple form of chemical heating. If something is burning, which is the definition of combustion, it is generating heat. This is why a fire tends to sustain itself. In decomposition, heat is generated by decomposing materials. This process is usually much slower than combustion. Like combustion, it requires an external source of initial heat to start the process. This type of heating usually becomes a problem only in the bulk storage of materials.

Spontaneous heating is very similar to heating caused by decomposition except that no external heat is necessary. With certain materials, the rate of oxidation at normal room temperature can become rapid enough to start open combustion. Linseed-oil-soaked cotton rags are a prime example of this potential.

When any material is dissolved in a liquid to form a solution, a certain amount of heat is usually generated. Although the heat created is typically not enough to start fires, some materials, such as sulfuric acid, can produce sufficient heat this way to become a potential ignition source.

Nuclear heating refers to the heat generated by splitting (fission) or joining (fusion) of two or more atomic nuclei.

MEASURES OF FLAMMABILITY

Several terms are important when evaluating the flammability of a material.

The flashpoint is the temperature at which a liquid gives off sufficient vapors for an external ignition source to cause a flame to flash across the surface of the liquid. However, if the ignition source is removed, the flame will go out because self-sustained combustion is not possible at this temperature. Several methods are available for testing flash point; Tag Closed Cup ASTM D-56, Tag Open Cup ASTM D-1310, Cleveland Open Cup ASTM D-92, and Pensky-Martens Closed Cup ASTM D-93 (Figure 2.7).

FIGURE 2.7 Pensky-Martens flashpoint tester.

The ignition temperature or firepoint is the temperature at which the material will begin self-sustained combustion if an external ignition source is used to initiate the process. This temperature is usually only slightly higher than the flashpoint. The auto-ignition temperature is the point at which the material has been sufficiently heated for combustion to occur without an external ignition source.

The flammable or explosive range identifies the percentage mixture of flammable vapor or gas in air that can be ignited. The flammable range is the area between the upper (UFL) and lower (LFL) flammable limits, also referred to as explosive limits (UEL and LEL). Gasoline, for example, has a lower flammable limit of approximately 1.5 and an upper flammable limit of approximately 7.5. This means that if its vapors are mixed in the surrounding air between 1.5 and 7.5%, and an ignition source is introduced, it will burn or explode. If the percent of vapors in air were 1%, the mixture would be too lean to burn because sufficient fuel would not be present. If the percent of vapors in air were 10%, the mixture would be too rich to burn because there would be too much fuel relative to the oxygen (Figure 2.8). Figure 2.9 illustrates a comparison of flammable ranges for several common substances.

A solid material's contribution to a fire is most commonly measured by its ease of ignition, flame spread, and smoke production. For testing and evaluation, solid materials are usually grouped into two primary categories: flexible solids, which include upholstery, furniture cushions, and clothing, and structural solids, which include solid building materials whether they are used in the structure or the contents.

Ease of ignition is tested to provide information about how much and how long heat must be applied to ignite the material under consideration. Flame spread

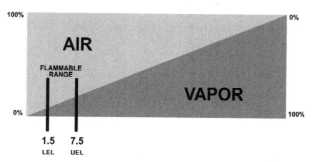

FIGURE 2.8 Flammable limits.

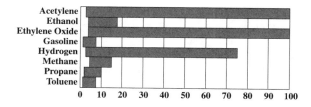

FIGURE 2.9 Flammable range comparison.

addresses the speed at which a fire, once ignited, will travel across the surface of the material. Flame spread testing is typically done in a Steiner Tunnel. One example of a test standard is NFPA 255 *Standard Method of Test of Surface Burning Characteristics of Building Materials*. Smoke production has traditionally been evaluated based on the amount of visible smoke and not on the chemical composition of the smoke. The current trend is toward more accurate measurement of the smoke's toxic components that may produce detrimental effects on people in a fire situation.

CHARACTERISTICS OF FLAMMABLES

Vapor pressure is the quantified description of a liquid's ability to release vapors (Figure 2.10). Atmospheric pressure is a downward force exerted by the atmosphere on the surface of a liquid; the vapor pressure is a measure of the opposing force exerted by the vapor pushing upward from the surface of the liquid. These pressures may be measured using a variety of units; atmospheres (atm), millimeters of mercury (mmHg), and pounds per square inch (psi) are a few examples.

The atmospheric pressure is relatively constant at any given point above sea level; the sea level value is 760 mmHg. Variation will occur at different elevations. The atmospheric pressure is lower in the mountains than it is at the beach. The pressure released by the liquid is a function of that liquid's characteristic. For example, kerosene has a vapor pressure of 5 mmHg at 100°F (37.7°C), which indicates that it will release very little vapor at normal temperatures. Toluene, a common solvent, has a vapor pressure of 21 mmHg at 68°F (20°C), which indicates it will release considerably more vapors than kerosene. Ethyl acetate, another common solvent, has a vapor pressure of 73 mmHg at 68°F (20°C), indicating that it will release more vapors than toluene.

Vapor density is a comparative measure with air always having a value of one. A vapor density less than one indicates the vapor or gas being considered is lighter than air and will tend to rise and dissipate. Vapor density greater than one indicates the vapor or gas being considered is heavier than air and will tend to sink and seek low points. Figure 2.11 illustrates this concept.

Most common flammable liquid vapors and gases are heavier than air. For example, propane has a vapor density of 1.6, indicating it is significantly heavier than air. Acetylene has a vapor density of 0.907, indicating it is slightly lighter than air.

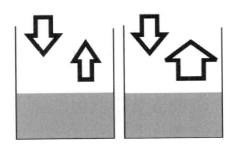

FIGURE 2.10 Vapor pressure.

FIGURE 2.11 Vapor density.

Specific gravity (also sometimes referred to as specific density) is another relative measure where water is always equal to one. This measure compares the density of a liquid with the density of water. A specific gravity of less than one indicates the liquid is lighter than water. A specific gravity greater than one indicates the liquid is heavier than water. Figure 2.12 illustrates this concept.

Solubility is the ability of the liquid to combine with water. Solubility is a scale, not a fixed point. Liquids may be insoluble, partially soluble, or completely soluble. Hydrocarbon liquids, such as gasoline, are not soluble. This information combined with the fact that the specific gravity of gasoline is less than one indicates that gasoline would float on the surface of water and retain its characteristics. Isopropyl alcohol also has a specific gravity less than one but it is soluble, so if it were released into water it would initially float but then form a solution with the water.

HEAT TRANSFER

Heat transfer is what allows a fire to spread. It plays a major role in our loss control efforts because understanding how a fire spreads increases our ability to prevent that spread.

Fire travels through four methods of heat transfer: direct flame contact, convection, radiation, and conduction. Direct flame contact, as the name implies, is the movement of the fire from one area to another by direct contact with flame. Direct flame contact is responsible for the initial spread of the fire during its incipient stage.

Convection is the movement of heat through a fluid medium such as the air and is the primary method of fire spread during the more developed stages of the fire. This method of heat transfer can move large amounts of heat over substantial distances within the structure. Convection currents tend to rise because the hot fire gases

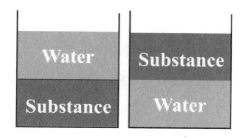

FIGURE 2.12 Specific gravity.

FIGURE 2.13 Convection.

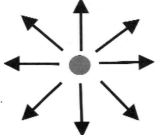

FIGURE 2.14 Radiation point source.

are lighter than the cooler surrounding air (Figure 2.13). If this upward movement is blocked, the currents will move horizontally. If that path also becomes blocked, the convection currents begin to bank downward. When downward banking occurs in an area, we refer to it as mushrooming, and eventually the entire area will fill with smoke, heat, and fire gases.

Radiation is the transfer of heat by way of energy waves. This transfer occurs equally in all directions and is not affected by air currents or transparent solid objects such as window glass. This method of heat transfer can make fires seem to jump from area to area or to ignite separate structures. Radiation's impact on fire spread will vary depending upon the source of the radiation. A point source projects energy equally in all directions (Figure 2.14). This tends to limit the energy striking any single object. A long source of radiation (Figure 2.15) tends to bombard a receiving surface with concentrated energy. A practical example of this type of fire spread is in a warehouse. If one stack of materials is on fire, the radiant energy from this fire will tend to ignite an adjacent stack.

FIGURE 2.15 Radiation long source.

Conduction is the transfer of heat through a solid object. This is not a very common method of fire spread but presents unique problems when it does occur. Conduction could, through piping, for example, start a fire in combustible materials on the opposite side of a solid wall. An example that most people are familiar with that illustrates this point is the handle of a pot on the stove becoming hot.

FLASHOVER

Flashover is a powerful phenomenon of fire behavior. A fire in an enclosed space, such as a room, will build up heat as it grows, and if no control measures are applied, the heat will reach the ignition temperature of most of the materials in the room. When this point is reached, the room will flashover. Most of the combustible materials in the room ignite and begin to burn in the same instant. This explosive fire development will fully envelop the interior of the room.

BACKDRAFT

Another unusual aspect of fire development that happens occasionally is a backdraft or smoke explosion. As a fire develops in an enclosed space, it consumes the available oxygen and generates carbon monoxide and heat. Carbon monoxide is a flammable gas. This situation provides two elements of the fire triangle in large quantities: the fuel of carbon monoxide and the heat from the ongoing fire. If air is introduced to this situation, the carbon monoxide will burn rapidly. This condition can create explosive forces. Proper fire control techniques can be used by the fire department to minimize the potential for backdraft.

PRODUCTS OF COMBUSTION

There are four basic products of combustion: flame, heat, smoke, and fire gases. Each of the four presents its own hazards to both a structure and its contents, including people.

Flames, which are the visible light mentioned in our definition of fire, are usually present. They are the most noticeable and threatening product of combustion. They are also probably the least significant as a real threat.

All fires produce heat, but the amount of heat produced varies based on what is burning and the conditions under which that burning occurs. In the area of fire growth and spread, heat is the most significant problem. If temperatures reach a high enough level, which is common in structural fires, heat can be sufficient to kill instantly by damaging the victim's lungs. Heat alone can be a major cause of fire-related damage even in areas the flames never reach.

Smoke is visible particles suspended in the air. Smoke can obscure vision and make escape difficult for individuals in the area of the fire. It can also damage the structure and its contents.

Fire gases make up a broad category that includes many individual gases produced by combustion. These invisible gases are the largest cause of fire deaths. They are usually undetectable by human senses and are present in hazardous quantities during almost all fires.

Carbon dioxide is present in all fires. Even "perfect" combustion under laboratory conditions produces this gas. An example of perfect combustion is:

$$\text{Methane} + \text{Oxygen} \rightarrow \text{Carbon dioxide} + \text{Water}$$
$$CH_4 + 2O_2 \rightarrow CO_2 + 2H_2O$$

It is also the gas we exhale, produced in our internal version of combustion. Carbon dioxide can increase the rate and depth of breathing in the individual exposed to it, causing hyperventilation. This elevated rate of breathing increases the exposure to the other fire gases and can cause dizziness, fainting, and headache. All of these affect the individual's ability to escape from the fire situation.

Carbon monoxide is credited with the vast majority of fire deaths. What is referred to as "died of smoke inhalation" in the evening paper usually means that a fire victim died from inhaling too much carbon monoxide. Carbon monoxide is produced when something burns in less than ideal conditions, which would be the case in all accidental fires. When we have a less than ideal concentration of oxygen in the area of combustion, the formation of carbon dioxide becomes chemically impossible, and the fire produces carbon monoxide. The red blood cells in the body, which carry oxygen to all parts of the body, have a greater affinity for carbon monoxide than oxygen. That means, given the choice, they will replace the oxygen they carry with carbon monoxide. Carbon monoxide affects the individual's ability to escape because the lack of oxygen in the blood interferes with all body functions. Confusion and disorientation may result as will, eventually, unconsciousness. Carbon monoxide is also flammable, a fact which was discussed under backdrafts.

Many other fire gases, such as hydrogen cyanide, sulfur dioxide, nitrogen dioxide, and acrolein, may be present depending on the materials involved in the fire. Many of these fire gases are individually more toxic than the two we discussed, but they are usually produced in much smaller quantities. Since the two main problems are carbon dioxide and carbon monoxide, we will limit our coverage of the others.

One final factor of possible importance is reduced oxygen content in the atmosphere surrounding the fire. As the fire burns it consumes oxygen. Heated fire gases displace oxygen, which creates two additional problems. First, the reduced oxygen level is obviously a hazard to people, since we breathe oxygen from the air. Second, reducing the oxygen causes a corresponding increase in the production of carbon monoxide — the more dangerous of the two fire gases we discussed.

EXTINGUISHING METHODS

All of the methods we use to control and extinguish fires are based on the fire square. We focus on removing one or more of the elements that allow the fire to continue.

The most common extinguishing method is to remove the heat. This is usually done with water. The goal is to apply water to the burning materials in sufficient volume such that the water is absorbing more heat than the fire is generating. If this is done, the burning materials will be cooled enough to drop below their ignition temperature, and the fire will go out.

Breaking down the chemical reaction is probably the second most common extinguishing method. Dry chemical, halon, and some halon replacement extinguishing agents function this way. They inhibit the ability of the materials involved to sustain a chemical chain reaction. If this reaction cannot be maintained, the fire will go out. Removing the oxygen is also a relatively common method of extinguishment. Carbon dioxide agents work this way, as do simple techniques like putting a lid on a pan. They work because they remove oxygen.

The last method, which is usually confined to special types of fires, is removing the fuel. One example is a fire involving a flammable gas. The ideal way to extinguish that kind of fire is to shut off the gas flow. Another example is the way most forest fires are controlled. A "break" created in the path of the fire removes the fuel and stops the fire's progress.

EXTINGUISHING AGENTS

WATER

Water is by far the most commonly used and readily available extinguishing agent. It is used in portable fire extinguishers, manual hose lines, installed systems, and as a base for foam-water systems. Water works well because it has a large capacity for absorbing heat. As we mentioned earlier, absorbing the heat cools the burning material to below its ignition temperature, thus causing the fire to go out. Water absorbs the most heat during its conversion to steam. For example, a pound of water at a room temperature of 70°F (21°C) will absorb approximately 150 Btu (158 kilojoules) to raise the temperature to 212°F (100°C), the boiling point of water. When water is vaporized into steam, it absorbs approximately another 970 Btu (1023 kilojoules). It is during the conversion to steam that the maximum benefits from the application of water are gained. Vaporization during effective use of water also reduces water damage. Under actual fire fighting conditions, it is not possible to get 100% vaporization or even come very close to it, but the goal is to apply water in a way that will provide the highest possible vaporization.

The remainder of the agents we are going to discuss are not ranked in any particular order. They are all specialized to some extent, and each one fills a specific need. Some of them are useful on more than one class of fire, but none of them equals the common availability or cost effectiveness of water.

DRY CHEMICALS

Dry chemical agents are fine powders (about the consistency of talcum powder) that are based on several chemical compounds. They are available in two categories: regular dry chemical agents which may be used on class B and C fires, and multipurpose dry chemical agents for use on class A, B, and C fires. They both function primarily by interrupting the chemical chain reaction. The multipurpose dry chemicals are compounds that allow the agent to adhere to surfaces, which is why they are effective on class A fires. Unless the class A rating is needed, regular dry chemicals should be used in class B and C hazard areas. The adhesion mentioned previously is a disadvantage in equipment and machinery fires because it makes cleanup much more difficult. Even with the regular dry chemical agents, the clean-up and salvage problem is a major disadvantage. This agent class generally provides the most rapid knockdown of flammable liquid fires available. Dry chemicals are used in portable fire extinguishers, wheeled fire extinguishers, vehicle and stationary hose systems, and local application installed systems both in structures and mounted on vehicles. Used by a trained individual, dry chemicals offer a potent fire-fighting agent for flammable liquid fires.

The most common regular dry chemical agent is sodium bicarbonate, baking soda. The most common multipurpose dry chemical agent is mono-ammonium phosphate. A common dry chemical used primarily for protecting significant flammable liquid exposures is potassium bicarbonate, typically referred to as Purple K. Purple K is approximately twice as effective, pound for pound, on flammable liquid fires as a regular dry chemical.

HALONS

Halogenated hydrocarbon agents, usually referred to as halons, are a group of gaseous agents that are effective in fire control. The two most common halons used for fire control are 1211 and 1301. Halon 1211 is typically used in portable fire extinguishers, and 1301 is normally used in installed systems. The halon agents extinguish fires primarily by interrupting the chemical chain reaction. Their major advantage is that they leave no residue, which makes them especially suited to computer and delicate equipment protection. The smaller units, those under seventeen pounds, are only rated for class B and C fires. The larger units, seventeen pounds and over, are rated for class A, B, and C fires. Halon is stored under pressure as a liquid. When discharged, it rapidly vaporizes to a gas. This behavior is referred to as a vaporizing liquid. Another advantage these agents offer is their holding ability. If a room is filled to the proper concentration with halon, usually about 7%, a fire cannot burn as long as that concentration is maintained. Halon ranks between dry chemical and carbon dioxide with respect to speed of fire control. The main disadvantage of halon is its environmental impact and cost. It is the most expensive of the agents we will discuss. Halon is also one of the chemicals connected with the depletion of the ozone layer. Its production has stopped in the United States, but it remains an approved agent in existing systems. Under the Clean Air Act (CAA), the U.S. banned the production and import of virgin halons 1211, 1301, and 2402 beginning January 1, 1994 in compliance with the Montreal Protocol on substances that deplete the ozone layer. Recycled halon and inventories produced before January 1, 1994 are now the only sources of supply.

This agent is used in portable fire extinguishers, wheeled fire extinguishers, vehicle and stationary hose systems, local application installed systems both in structures and mounted on vehicles, and total flooding installed systems.

CARBON DIOXIDE

Carbon dioxide (CO_2) is a gaseous fire control agent that is stored under pressure as a liquid. It is rated for class B and C fires. The major advantages of carbon dioxide are that it leaves no residue and is nonconducting. It does not afford as much fire control capability as halon or some halon substitutes, but it is much less costly. It functions by excluding oxygen from the fire.

The main disadvantage of carbon dioxide is that it can create an oxygen deficient area where it has been used that poses a significant risk to personnel. This agent is used in portable fire extinguishers, wheeled fire extinguishers, vehicle and stationary hose systems, local application installed systems both in structures and mounted on vehicles, and total flooding installed systems.

FOAM

Foam is a general category of extinguishing agents that includes a wide variety of specific foams for special purpose application. Foam is used in portable fire extinguishers, wheeled extinguishers, manual hose lines, fixed hose systems, and a variety of installed systems. Two major types of foam are chemical and mechanical. Chemical foams are created by a chemical reaction and are rarely used. Mechanical foams are created by mixing foam concentrate with a specific proportion of water to form a foam solution. Several types of proportioning devices are used. One of the most common is the in-line eductor. The pickup tube is placed in the foam concentrate, and water flows through the eductor. The passing water creates a venturi, which draws the foam concentrate into the stream. The metering valve controls the percentage of concentrate to ensure a proper mixture. The foam solution flows through the hose to a nozzle. Air is introduced to the foam solution at the nozzle in a process called aeration to form the finished foam. The finished foam is a bubbly substance that is similar to soap suds in appearance.

Foam is suitable for use on class A and B fires, but is specifically designed for class B hazards. Foam involves several of the extinguishing methods we discussed. It is mostly water, so it offers a cooling capability. It is designed to float on the surface of a flammable liquid, forming a barrier between the fuel surface and the air, so it excludes oxygen. This barrier acts to remove the fuel from the fire situation by sealing it.

All mechanical foams have two basic features that must be considered: the proportioning percentage and the expansion ratio. The proportioning percentage is the percent by volume of foam concentrate to water. The most common percentage is 3, and this type of foam may be referred to as 3% foam. Foam is produced in 1, 3, 6, and 9% types. A single variety of foam is often available in more than 1% type. The expansion ratio of the foam is the ratio of expansion that occurs when the foam is aerated. An expansion ratio of 10:1 indicates that for each cubic foot 28.3 liters of foam solution, 10 cubic feet (283 liters) of finished foam will be created. The main categories of foam expansion ratios are low ratios at 10:1 and high ratios from 100:1 to 1000:1.

Several characteristics are common to all foams and are used as criteria for comparing their quality. Foam must be able to flow freely over the surface of a liquid. This flow ability enables the foam to spread over the liquid's surface and aids in achieving rapid, complete coverage of the liquid. The foam should provide a stable blanket once applied. Foam will begin to lose water as soon as it is applied, and the speed at which the water is lost is the foam's drainage or drain down rate. The slower this drain down rate, the longer the foam blanket will last. Since the foam is used as a fire control agent, heat resistance is an important characteristic. The foam should be able to withstand the heat from a fire and hot surfaces in the area. This ability is sometimes referred to as burnback resistance. More specifically, it is the capability of the foam to maintain an effective blanket in the proximity of a fire. The foam should also resist contamination from the product to which it is applied. This is referred to as fuel pickup. Several specific types of foams are used. Each type of foam has advantages and disadvantages when compared with the others. The nature of the hazard being protected is the major factor when deciding on the specific foam that will be most effective.

Protein foam, which is made of natural materials, is one of the oldest types of mechanical foam. It offers high heat resistance and creates a cohesive, stable blanket. It is not as free flowing as the newer types of foam and, therefore, does not control fire as rapidly. Protein foam is no longer frequently used due to the improvements made in synthetic foams. It is most suitable for situations in which high heat resistance and stability over longer periods of time are important considerations. It is also not compatible with dry chemical agents.

Aqueous film forming foam (AFFF) is a free flowing foam that provides rapid fire control. It is the most common type of foam in general use and can be employed effectively on a wide variety of fire situations. It provides more rapid fire control in most situations than the other types of foam. Another major advantage of AFFF is its compatibility with dry chemical agents. This allows the use of these agents in combination, typically called twin-agent systems, which increases the effectiveness of each. Dry chemicals offer more rapid fire knock-down than foam, and foam can secure the fuel surface against reigniting. It does not require application with special foam equipment.

Alcohol-type foam is very similar to AFFF except that it is designed for use on polar solvent fires. These materials are water soluble liquids such as alcohol, methyl ethyl ketone, and other common solvents, which would destroy the effectiveness of ordinary AFFF. Alcohol-type foams have additional chemicals that cause a membrane to form on the surface of the solvent which prevents the solvent from breaking down the foam. These types of foam may be used on ordinary flammable liquid fires.

Fluoroprotein foam has some of the advantages found in both protein foam and AFFF. It has higher heat resistance than AFFF and resists fuel pickup more effectively. It is more free flowing than protein foam, and it is ideal for subsurface injection, which is a technique used to extinguish large flammable liquid storage tank fires. It is also compatible with dry chemical agents.

High expansion foam is most suited for extinguishing fires in confined spaces within structures. This foam will also displace the heat and smoke in the fire area if sufficient ventilation is provided during the application.

DRY POWDER

Dry powder agents are designed to control fires in combustible metals (class D). The two most common agents in this category are G-1 and Met-L-X. Dry powders function by creating a crust over the surface of the burning metal. To extinguish the fire, this crust must completely cover the burning surface. These agents are usually applied by hand scoop or portable fire extinguisher. Graphite and sodium chloride are two common examples of these agents.

WET CHEMICALS

Wet chemical agents, typically potassium acetate, are designed specifically for cooking oil and grease fires. They are available in portable fire extinguishers and installed systems.

HALON SUBSTITUTES/REPLACEMENTS

Due to the environmental issues associated with halon, a number of substitutes and replacements have come onto the market in recent years.

INERGEN® is manufactured by Ansul®, Inc. and is a mixture of nitrogen, argon, and carbon dioxide. Normal atmospheric air contains approximately 21% oxygen and 1% carbon dioxide. This agent reduces the oxygen content in the protected area to approximately 12.5% and increases the carbon dioxide to approximately 4%. Combustion with an open flame requires approximately 15% oxygen. Human respiration normally requires approximately 16% oxygen. The increased concentration of carbon dioxide stimulates human respiration though and permits normal breathing in the reduced oxygen concentration.

FM-200® (1,1,1,2,3,3,3-heptafluoropropane) is manufactured by the Great Lakes Chemical Corporation and is another substitute for halon. FE-36™ (1,1,1,3,3,3-hexafluoropropane) manufactured by DuPont® is another halon replacement. It is aimed at replacing Halon 1211, which was used primarily in portable fire extinguishers. It may also be used in explosion suppression systems.

3   Loss Control Programs

CHAPTER OBJECTIVES

You will be able to identify and explain:

- The reasons loss control programs are necessary
- Major components of a loss control program
- Major approaches to loss prevention and control
- The loss control process

You will also be able to:

- Develop a loss control program
- Organize a loss control program
- Implement a loss control program
- Manage a loss control program
- Evaluate a loss control program

WHAT IS A FIRE LOSS CONTROL PROGRAM?

Fire loss control programs are specifically focused efforts aimed at reducing the risk of loss, the magnitude of potential losses, and the impact of those losses on continuing operations. They are based on the accurate assumption that it is more effective and ultimately less costly to manage the loss potential of an operation prior to a loss rather than deal with the impact after a loss has occurred.

WHY ARE LOSS CONTROL PROGRAMS NEEDED?

Loss control programs are needed for a variety of reasons. As individuals responsible for some aspect of a loss control program, we mistakenly tend to assume that the need for the program is obvious. Some managers see loss control only as a means of complying with compulsory regulations, while other managers perceive it as something that must be done to appease the insurance company. Still others view it simply as a method for meeting corporate standards. The majority of line managers see loss control as more of a cost than a benefit.

Educating people whose perspectives may be different necessitates a planned explanation which is similar to a sales presentation. The ability to do this effectively is critical to success in loss prevention and control. Whenever approval is gained to start a project or to purchase equipment or materials, a sale has been made.

A key point that must be instilled in the minds of managers is that loss control makes good business sense. If loss prevention and control programs did not provide benefits to the business, they would not exist. Management personnel must be convinced of this important point. Once management understands the positive impact of loss prevention and control on the successful operation and profitability of the organization, gaining approval for specific projects and activities becomes easier.

WHO NEEDS A LOSS CONTROL PROGRAM?

All organizations need some form of loss control program. Even in the home, loss prevention and control are important. The purchase of a fire extinguisher, installation of a smoke detector, and storage of gasoline for the lawn mower in a safety can are all examples of loss prevention and control efforts at the household level. There is not a single organization that does not need some type of loss prevention and control program. People create many excuses for not having a loss control program. Some think it is too expensive. Others believe that it requires more time than it is worth. Some feel it is a waste of effort.

All of these excuses stem from the failure to understand the basic purpose of loss prevention and control and the proven effectiveness which efforts in this area have demonstrated in the past. The people making excuses instead of progress fail to recognize that loss prevention and control are critical issues that must be addressed by any organization seeking long-term success. These programs do cost money, take time, and require effort. It has been proven, however, that the money, time, and effort spent do pay. In the long run, loss prevention and control always cost less and require less time than losses.

No matter what size organization, what is being produced, what service is being provided, where the organization is located, or any of a hundred other factors, a loss prevention and control program is essential. The size, scope, and complexity of the program will vary considerably from one organization to another, but the need for a program is always present.

LOSS CONTROL REQUIREMENTS

U. S. LAWS AND REGULATIONS

The Occupational Safety and Health Administration (OSHA) is the primary federal regulatory agency involved in fire loss prevention and control in the United States. These regulations focus primarily upon protecting people in the context of their employment. They are not aimed at protecting the public, though some benefit may result to public safety as a spin-off of protecting employees. They do not care if property is lost as long as no employees are injured or killed.

STATE AND LOCAL LAWS AND ORDINANCES

State and local level governmental bodies are primarily focused on the protection of all people and secondarily on property protection. The property protection elements are included to minimize the impact of loss on the community. For example, if no laws existed to help limit the number and size of fires, communities would have to invest more heavily in fire departments and other resources to handle these fires. That would result in more taxes being imposed on everyone in the community.

Requirements for fire loss prevention and control at the local level are determined and enforced by the authority having jurisdiction (AHJ). Who fills this role will vary from one location to another. It may be your local fire department or a local code enforcement officer. It may be a state level fire marshal or department of insurance officer. There may not be anyone in your location that is specifically designated to enforce fire requirements. This lack of enforcement presence should not lead you to assume there are no requirements.

Standards organizations covered in the next section often yield a significant amount of power to the AHJ. They may interpret and apply requirements locally with a relatively free hand.

NATIONALLY RECOGNIZED AND CONSENSUS STANDARDS

Consensus standards are developed by committees that are intended to represent all of the interest groups that will be affected by the standard. Organizations have rules governing participation on committees and balanced representation.

Many organizations develop and publish nationally recognized and consensus standards. Some of the major organizations in the United States include:

- National Fire Protection Association
- Building Officials and Code Administrators
- American National Standards Institute
- American Society for Testing and Materials
- American Society of Mechanical Engineers
- Underwriters' Laboratories
- Factory Mutual

These standards represent minimum accepted practice. They may or may not be required by law depending on your local authority having jurisdiction. Many of the early OSHA regulations were taken directly from existing consensus standards at the time OSHA began in the early 1970s.

Recognized testing laboratories are a subset of these standards organizations. The two most common in the fire protection field are Underwriters' Laboratories and Factory Mutual. These organizations go a step beyond the development of standards and actually test products and devices against the requirements of their standards. Products that meet requirements are listed as approved for their intended use. These products are allowed to use approval marks (Figures 3.1 and 3.2).

FIGURE 3.1 UL symbol.

FIGURE 3.2 FM symbol.

INSURANCE COMPANY REQUIREMENTS

Insurance company requirements are focused exclusively on preventing dollar losses that they may have to pay out to an insured party. They are not there to help you. They are there to protect their investment. The objective of the relationship from the insurance company's perspective is to collect more premium dollars from their clients than they have to pay out in reimbursement for losses. If they succeed, they make money; if not, they lose money. They are generally not concerned about the protection of people unless they provide workers' compensation or liability insurance. They have no motivation to consider the overall cost-effectiveness of the property protection approaches they propose. Their clients, however, have an essential need to evaluate those issues.

These comments are not intended to be anti-insurance. Insurance fulfills a critical role in loss prevention and control efforts. It provides a safety net for business survival in the face of catastrophic losses. It is, however, important to remember their role and perspective when working with them. It is not to a company's advantage to blindly follow all of their protection recommendations.

LOSS CONTROL APPROACHES

The basic approach to loss prevention and control programs can be a major factor in the success of the program in terms of the results obtained compared with the resources expended. The main purpose of this discussion of basic approaches to loss control is to establish that the focus of loss prevention and control must be on human behavior. Many techniques and approaches can be useful in preventing and reducing losses, but the focus and central theme of loss control needs to be people — their attitudes, behaviors, actions, and inactions. Two basic approaches to loss prevention and control can be taken: analytical and behavioral.

An analytical approach tries to consolidate all factors involved in losses into some quantifiable form and focuses primarily on complex systems and engineering solutions. This approach is unsatisfactory in several ways because it assumes that a number can be attached to each factor involved in a loss. This is not practical for several reasons. For example, what numerical value can be assigned to a human life? Even if it is accepted in principle that a dollar value can be coldly attached to a human life, this determination only scratches the surface of the problem. How should this dollar value be established? Should it be based on the individual's earning power, contribution to company earnings, cost of medical expenses, value of life insurance, or any of a long list of possible choices? Even a generalized attempt to determine this value becomes extremely complicated and open to highly emotional debate.

Even the less emotional factors become complex. Human error needs to be factored into a loss control program. Yet human error cannot be accurately quantified. It is possible to derive a number that represents human error under some very strict hypothetical conditions. Although this number may be a fair representation of what happens under the hypothetical conditions, these calculations become an exercise in paperwork when the actual conditions are compared with the hypothetical ones.

The final issue deals with what this approach accomplishes even if the numbers are fairly accurate. They would constitute the foundation of an excellent technical paper for publication in a trade journal, and that's about it. We would be no closer to preventing losses or reducing their consequences.

Anyone interested in reducing losses and their consequences should maintain records to track progress. Statistics are necessary to help identify areas of a program that need improvement. However, the focus should be on the progress, not on the numbers in the records. The numbers should be used to measure the results of a program; they should not become a goal in and of themselves.

The analytical approach to loss control also focuses on systems and engineering. The goal, taken to the extreme, is to design systems that prevent all losses. The underlying assumption necessary to give primary emphasis to systems and engineering is that the physical environment can be shaped to remove the loss-creating potential. This is the fundamental assumption made in fail-safe engineering: engineering something in a way that if it fails, it always fails to the safest mode. It is a sound method but there are definite limits. By placing the emphasis on analysis and engineering, attention is diverted from the area with the greatest potential for loss reduction: human behavior.

The other major approach to loss control focuses on people and their behavior. To be successful in preventing losses, our main impact must be on human attitudes, beliefs, and behaviors. This focus on the human element is not restricted to the front-line employees. Behavior issues up to the top management level of the organization have an impact on loss prevention and control.

When analyzing almost any loss situation, one discovers that the initiation of the loss, failure to prevent the loss, expansion of the loss, or all three can often be traced to people. Somewhere in the process of the loss scenario, an individual either did something that should not have been done or failed to do something that should have been done. Even safety experts sometimes do something or fail to do something and later realize that they broke one of their own safety rules.

Frequently, the investigation to discover the cause of a loss stops too soon. For example, a determination is made that an overheated bearing in a motor caused the fire, and that is where the cause determination investigation usually stops. The fire report indicates the fire was caused by an equipment malfunction. But was it really? Was the preventive maintenance schedule checked to see if the motor was maintained properly? Was the bearing installed properly? The answers to these questions are usually not reflected in the fire investigation. This may seem like splitting hairs, and, to an extent, maybe it is. The important point is that people cause losses, not things.

Even in the rare cases when a direct human cause contribution cannot be found, human error is usually involved in expanding the damage caused by the problem. It is crucial to keep this concept in mind. Most of our efforts in loss control must be aimed at people and their behavior and attitudes.

The most effective approach to loss control focuses on the human factor. Both analysis and systems engineering are important, but their use should be aimed at dealing in a useful manner with human behavior. Analysis should focus on why losses occur and what can be done to prevent them. Analyzing to the extent that a formula is developed to statistically predict the probability of a certain loss scenario requires much effort and provides little useful information.

Engineering and systems methods must focus on compensating for human inadequacies. The effort required to engineer all loss potential out of a process or system is extensive and usually cannot accomplish the objective. The development of systems which limit the potential for human error and lessen its impact is more effective. Systems methods offer benefits primarily by helping to ensure that the human factor is taken into consideration properly and to provide checklist-type guidance for loss prevention and control efforts.

This book will spend a great deal of time covering the technical aspects of loss control. While these technical approaches are essential, they should never cause us to lose sight of the fact that we must emphasize the human factor to achieve overall success in our loss control efforts.

Most of the technology used in loss control is designed to compensate for the human factor in some way. We will look at the way our techniques and methods impact on the human factor in loss control throughout this book.

LOSS CONTROL PROCESS

The loss prevention and control process illustrated in Figure 3.3 is designed to help evaluate the alternatives concerning loss control. It is not intended to be all-inclusive. The process provides a starting point which will help channel the user's thinking when considering loss control options.

The first step in the process involves an analysis of the loss potential. Two primary areas are evaluated initially: what can be lost and what may cause a loss. After these items have been identified, they are divided into controllable and uncontrollable loss factors. For example, the manufacturer of a product made with highly combustible materials that must be heated to be formed cannot eliminate the fuel or heat without making the process impossible. In these circumstances, the fuel and heat being together is an uncontrollable element and requires a search for other ways to reduce the loss potential of this operation. Even though the basic heat and fuel combination in this example is not controllable, certain parts of the situation are. This demonstrates that controllable and uncontrollable loss factors are not always mutually exclusive. A single operation or process may have certain characteristics that fall in the controllable category while others are considered uncontrollable.

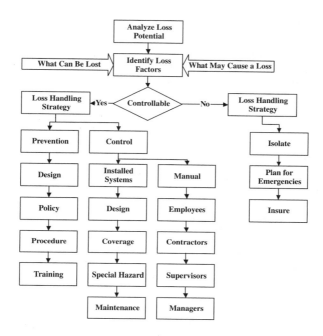

FIGURE 3.3 Loss control process.

The uncontrollable portions that are identified require a very different loss handling strategy than controllable elements. The uncontrollable factors are dealt with primarily through isolation of the hazard to the extent possible. This serves to limit the magnitude of a loss should one occur. Planning for emergencies should be more comprehensive under these circumstances because it is assumed that, at some point, an emergency is certain to occur. For that part of the risk which cannot be minimized, additional insurance against the loss is the last resort. The term uncontrollable, as used here, is not intended to imply that a factor is never controllable. It refers to the portion of the loss factors which cannot realistically be controlled at present. Many innovations in fire loss control have resulted from developments that have allowed control in an area where control was not previously possible. The invention of safety solvents, for example, changed where and how flammable liquids needed to be used in industrial processes. The introduction of the combustible safety solvents to replace flammable solvents in some uses improved the ability to control a factor involved in losses. In the manufacturing example used above, the invention of a noncombustible material with the same forming characteristics as the current material would make the loss control of the process more controllable. Uncontrolled factors are also not meant to indicate a complete lack of control in influencing losses. When uncontrollable factors are involved in a process, it simply means that more effort must be expended to adequately handle those factors that are controllable. The main reason for making this distinction is to assist in the allocation of resources. With the limited resources available for loss control, it is essential that our efforts be utilized in areas in which they will have the greatest impact.

Loss handling strategy for the controllable loss factors focuses on two primary tracks: prevention and control. Prevention involves the use of policies and procedures which minimize the potential causes of loss. A common example of a loss prevention policy is one that prohibits smoking in certain areas. Employee training in loss prevention can have a major impact on reducing the chances for a loss. An example of this is training employees for the proper dispensing and use of flammable liquids. Engineering also plays a significant role in reducing the occurrence of losses. For example, an automatic shut-off tied into a temperature sensor can eliminate the potential for overheating that could cause a fire.

Control as a loss handling strategy begins where prevention efforts have failed. The items considered under control (see Figure 3.3) are employed when a fire has started and are designed to minimize the damage and to control the fire as quickly as possible. Manual fire fighting is one of the first lines of defense. These fire fighting efforts can be made by employees, emergency teams, fire brigades, or public fire departments. A combination of several of these approaches may be used, but all of these options must be analyzed and evaluated.

Installed fire protection takes three primary forms. The first is systems which are intended to limit the growth and spread of the fire. Automatic fire doors and automatic fuel shutoff valves are examples of this type of system. Second, there are systems which act as sensor devices. An alarm system is the best example of this type. Although this kind of system does not contribute to the actual control of the fire, it enhances the overall response to the emergency by providing warning of the problem. The third category of systems includes those that actually exercise control over the fire. They discharge extinguishing agents directly on the fire and control or extinguish it without human intervention.

Planning is a control strategy that can be applied in conjunction with all other items. Using the process discussed here is, in part, a planning activity. Planning involves the evaluation of circumstances and anticipation of needed actions prior to an actual fire. The more accurately the nature of the fire emergency can be predicted, the better the response will be. Procedures are closely linked to planning. They establish a way of handling a situation in advance to ensure that all items that need to be dealt with during an emergency are, in fact, taken care of. Evacuation procedures are an example of the type of item that can be reduced to a rather fixed set of actions.

Employee training is an essential element of the control strategy. Every employee must be trained regarding the actions he or she must take during an emergency. This is true even if all the employees are expected to do is evacuate.

LOSS CONTROL PROGRAM ESSENTIALS

The essential elements of a loss control program are illustrated in Figure 3.4.

FIGURE 3.4 Loss control program elements.

MANAGEMENT COMMITMENT

The first element that must be present for a loss control program of any type to be effective is commitment on the part of management to having a loss control program that is aimed at getting results and not just at looking good on paper. This commitment must come in several forms. The most important one is also usually the most difficult to obtain. The organization's top management must recognize personally that loss control is an essential business priority. The fact that effective loss control is necessary not only from a human perspective but also in terms of cost/benefit seems fairly obvious. It is not, however, obvious to many people. Top management personnel often view loss prevention and control as a cost, not a benefit. One of the keys to achieving the appropriate commitment to a loss control program is convincing management that loss control is an integral part of the operation. Many managers have the impression that loss control is an add-on activity. The same individual who wouldn't dream of running his/her business operation without insurance cannot seem to grasp why money needs to be spent on training the workforce in the use of fire extinguishers. This gap in understanding must be bridged with education. As the individual with safety responsibilities in your operation, your first priority is to educate top management. Once top management understands the essential nature of loss control, you have already taken a major step toward meeting your other needs from top management.

Policies

The first specific item we need from top management is an effective written policy. This overall loss prevention and control policy must originate from the top management personnel of the organization. For a policy to be effective, it must be important to those who write it and understood and acted upon by those who read it. The individuals responsible for safety fall in the middle of this transaction. Top management will need advice on what policies to write and how they should be written, and employees will need training and assistance to use the policies effectively. Finally, the policy must be enforced.

Support

The second major area of top management involvement is support. Support from any manager usually takes two basic forms: time and money. Through their time, managers visibly demonstrate that the loss control program is important to them personally as well as to the organization. Through money, they authorize those responsible for safety to apply business resources to have an impact on losses. Either of these expenditures will have to be justified on a case-by-case basis. This is as it should be, but the general management philosophy concerning whether loss control efforts are worth an investment of time or money must be positive for the overall program to be successful. This does not mean that you will get approval for every project or idea you take to management. You will not. The most crucial element is that management is receptive to investing in loss control.

Example

Setting an example is a powerful communications tool. Top management will demonstrate their true commitment to loss control by the example they set. What management does will have a far greater impact on how the employees view their commitment to loss control than what they say. If the first time a loss control policy becomes inconvenient to management a temporary exemption is authorized, that policy will have lost credibility throughout the entire organization. Top management must resign themselves to following their own policies even when it would be easier to bend the rules. Although a policy occasionally needs to be bent, management must be careful. The negative impact of a situation in which a policy must be broken can be minimized by prior education. If an understanding is developed in the workforce as to why the policy has to be bent, stressing that it is not purely for the convenience of management, the situation can be dealt with and the policy can be left intact. The education must occur prior to the breach of policy, or it will not be effective.

An example of this might be the installation of a new piece of equipment. During the process of the installation, an emergency exit will have to be blocked for an extended period of time. It is essential that management communicate to the employees that work is occurring in this area, the temporary nature of the blockage, and why it could not be avoided. If this is not done, the next time an employee is criticized for blocking an exit with a pallet of materials, he/she may not understand why what they did was against policy but what management did was allowed.

Management must also set a personal example by always following the safety policies and using safe practices. Small things like not wearing safety glasses during a plant visit are noticed by employees and can have a major negative impact on the loss control program.

CLEARLY DEFINE AND ASSIGN RESPONSIBILITY

A general management principle that is particularly important to the overall loss control program states that management must define who is responsible for each aspect of the loss control program. Due to the nature of the loss control process, this can, at times, present difficulties. Often, the responsibility for loss control must be given to a number of individuals, each having specific areas of concern. This method works well as long as the individuals know which areas are their responsibility and understand what is involved with those responsibilities. Too often, this assignment is left to chance or made as an add-on assignment to someone who already has too much to do. Both these situations create the impression that loss control is not really a priority issue.

DELEGATE AUTHORITY

Delegating authority is also a general management principle. An individual cannot be held accountable for a task if sufficient authority to complete the task has not been provided. Sometimes a person is made responsible for loss control as an add-on task to his/her main job and has not been given any real authority to accomplish anything. This frustrates the individual and prevents any significant loss control efforts from being established.

ESTABLISH PROGRAM OBJECTIVES

The establishment of effective objectives is essential for producing results in an area. The principle of setting goals and objectives applies to any endeavor, including loss control. The loss control program must have objectives if it is to produce results. Our efforts need the focus provided by these objectives to gain maximum benefits. If we fail to plan our loss control efforts, we will expend resources haphazardly and not realize much of the benefit that could have been obtained. Top management must be involved in this process of establishing the objectives of the overall loss control program. This is important for several reasons. First, management support is vital. It is much easier to get this support from managers that have been involved in setting specific targets for the program. This process also forces safety specialists to look more objectively at the goals they have set for the loss control program. This self-analysis is essential to keeping their own thinking on track. Finally, this objective-setting process provides a way of measuring their degree of success. Without a specific objective, it is impossible to determine whether the desired result was accomplished.

Objectives should be as specific as possible. For example, the target may be to reduce the number of fires in the plant by 10%. A completion date should be set for all objectives. As part of the objective-setting process, the methods that can be used and the resources that will be needed to accomplish the objective should be evaluated. The objective then begins to take the shape of an action plan. By explaining in detail how the objective will be accomplished and what resources will be needed, support can be won from the management, who will immediately see what results are intended from the expenditure of the resources.

INVOLVE SUPERVISORS AND EMPLOYEES

For a loss control program to be successful on a long-term basis, it must have the support not only of top management, but also of intermediate level supervisors and all other employees. Without this support, some progress may be made, but it will always be more difficult and usually not long lasting.

There are several levels of employee participation in loss control. The lowest level is no involvement at all. At this level, employees perceive the loss control program as just another management harassment tactic. The program is viewed negatively, and the employees are uncooperative in all aspects of loss control. Fortunately, this is rare, but it does exist in some companies. Operating a loss control program in this environment is difficult at best. One of the first objectives of a program in this situation must be to educate the employees concerning the importance of loss control. This education program should stress the direct benefits to the employees. Often, the employee attitude toward loss control reflects deeper feelings about management in general. These feelings can be difficult to change by the individual responsible for safety. Usually, a change must first occur in the top management attitude toward working with the employees.

At the next level, the employees see the loss control program as a necessary evil. The perception is neither positive nor hostile. Safety will not get a great deal of cooperation, but most employees tolerate the loss control policies and, for the most part, follow them. There is still very little understanding among the employees that loss control provides direct benefits to them. They may begin to watch for hazards to some extent and are more conscious of them.

The next level of employee attitude toward loss control is when the employees have gained sufficient understanding of loss control that they have begun to see how loss control benefits them directly. This situation creates more cooperation and participation from the employees. They will begin to feel that it is their responsibility to report hazards they discover. The cooperation received from employees contributes to the overall loss control program.

The highest level of involvement by employees is active support and participation by all. The employees have a thorough understanding of the loss control program and the direct benefits to themselves from their participation. Cooperation is obtained easily, and employees actively seek ways to contribute to loss control efforts. This level is obviously the ideal in terms of loss control efforts.

Two specific techniques can help increase employee involvement: safety committees and a suggestion program. Safety committees not only offer the best method, but also serve a number of functions: (1) they provide highly visible evidence that loss control is important to the organization, (2) they offer a direct method for the individual employee to participate and have input into the process, and (3) management is provided with an excellent source of ideas for accomplishing loss control objectives.

It is important to establish the purpose and scope of the safety committee at the beginning of the program. All participants must understand the goals and objectives of the committee. Employees must have a clear understanding of the role of the safety committee in the overall loss control program. It should be impressed upon participants that the committee meetings are designed to generate ideas and foster positive change; they are not gripe sessions. Ground rules should be established for the meetings in advance. All meetings should have an agenda which is distributed to participants prior to the meeting. A time limit should be set for the meeting to avoid endless discussion.

Another specific technique is a suggestion program. Although this option does not accomplish the magnitude of results that a safety committee would, it is, at least, a start. It is essential that all employees be encouraged to make suggestions. All suggestions must receive a response. Thank employees for making suggestions. If a suggestion will be used, explain how it will be implemented and when. There should be a system to provide some type of incentive award to those making suggestions that are used. If a suggestion will not be used, explain why and encourage the individual to contribute again in the future.

EMPHASIZE LOSS CONTROL IN ALL AREAS OF OPERATION

Loss control must be a consistent effort which is reflected in all phases of an operation, not just the most convenient ones. Maximum positive results can only be achieved if the loss control efforts have an impact on all phases of the operation. This seems at first like a small point, but remember that loss control is not a part-time activity or one that can be applied successfully to only a few operations within an organization. A commitment to loss control must be made throughout the organization or the results will be less than ideal. Loss prevention and control must become an integral part of the way the organization operates.

PROVIDE ADEQUATE TRAINING

Training will be stressed many times in the various sections of this book. At this point, it is important to realize that training is one of the most essential elements in the overall success of the loss control program.

Personnel cannot be expected to perform their tasks safely if they have not been taught how to do so. This point should be so simple that it does not even need to be mentioned, but it is not always easily recognized. Problems can be encountered that a ten-minute briefing on proper procedures would have prevented. Too often, however, no time is spent giving instructions until a problem occurs. The first training challenge is to educate all personnel concerning the importance of loss control and their role and responsibilities in the overall loss control program.

Effective Inspection Program

The physical features of a loss control program must be thoroughly inspected on a regular basis. Procedures and their application must be reviewed frequently. The only way to ensure that things are as they should be is to check. Safety managers who spend more than 20% of their time at a desk are spending too much time on paperwork and not enough on application. Losses occur in every plant. That's where a safety manager should spend the most time. Inspections are also advantageous because they make the safety manager a highly visible figure in plant operations and indicate to all personnel that loss control is important. One of the weakest areas is when the individual responsible for loss control is only active in loss control part-time. If the personnel manager is also responsible for safety, he/she probably cannot take the time necessary to effectively inspect and spend time on the plant floor.

Investigate All Losses

To have an impact on the future we must understand the past. All losses are caused by something, and determining the cause of a loss is important to the prevention of another loss in the future. An investigation may also identify a need for improvement in our loss control policies or the enforcement of those policies. A valuable source of information is irretrievably gone if all losses are not fully investigated. Even small losses should be investigated completely. The loss may have been small this time, but the cause of the loss, if not corrected, may create a major loss next time. Investigations must take place as soon as possible after losses, or valuable evidence may be lost or damaged. An investigation also makes it possible to correct an inaccurate initial impression of the cause of the loss upon closer examination. A thorough investigation also provides information needed for insurance and other legal considerations.

Maintain Adequate Records

Effective records are essential to tracking the progress toward the established objectives. In addition, the maintenance of adequate records permits the safety personnel to gain valuable information from past losses that can be applied to prevent or more effectively control future losses. Trends can also be seen over time, and good records provide the necessary information to make adjustments to future loss control efforts.

Legal requirements often specify which documents are to be retained. Injury reports, training records, and many other records must be available to satisfy the requirements of regulatory agencies. Insurance companies also frequently stipulate that certain records be maintained.

Enforce Loss Control Policies Consistently

All policies developed for a loss control program are useless if they are not enforced properly. Policies will quickly lose their impact if employees perceive that those policies will not be executed. Even worse than not enforcing policies is to enforce them inconsistently. Loss control policies cannot be effectively carried out on a hit-or-miss basis. Enforcement is a full-time effort. Consistency in distribution of rewards and punishments is also critical. If employees perceive that a policy has been applied unequally, based on who broke the policy or where the infraction occurred, the policy will lose most of its effectiveness.

PLAN FOR EMERGENCIES

Planning for emergencies is essential, and it is crucial that time is allotted for this activity. No matter how effective the loss control program may be, an emergency is going to occur eventually. How much is lost during that emergency is going to depend largely on the extent and quality of the prior emergency planning.

IMPORTANCE OF MAJOR LOSS CONTROL PROGRAM COMPONENTS

The factors that have just been discussed are the major components of a successful loss control program. Although some success can be attained without addressing all of these elements, lasting success requires that they all be utilized. The amount of commitment to any individual component will vary considerably from one organization to another. If, for example, an organization manufactures toxic chemicals, it will need a more extensive emergency planning effort than a manufacturer of non-hazardous products. All of the components must be addressed by all organizations.

ORGANIZING FOR LOSS CONTROL

Loss control programs are very dependent on their organization for success. A well-organized loss control effort is much more likely to be successful than one that is not effectively organized. For the purposes of our discussion, we are going to look at loss control organization in two separate phases: non-emergency and emergency. In practice, these phases have to mesh to form a coordinated organization. They are presented here individually, however, to make the description easier to follow. The relationships and interactions discussed herein are necessary no matter what size organization is involved. In a small organization, the relationships and interactions among personnel tend to be less complex than in a large one. There will tend to be less chance of overlooking one of the organizational functions in a large organization due to the additional personnel. Your personnel titles will probably differ from the ones used in these examples. However, the focus should be on the functions and interrelationships of the personnel and not on their specific titles.

Both internal and external personnel will be considered because both have a major impact on any loss control program. Most of the external personnel that are included in this discussion of organizing for loss control have a vested interest in the success of our efforts. This makes them valuable allies.

NONEMERGENCY ORGANIZATION

Figure 3.5 depicts a possible organizational structure for day-to-day, nonemergency operations in a larger organization. We begin with top management whose direct daily involvement will usually be limited. In our example, we are assuming a multi-facility corporation in which there is usually a corporate-level loss control manager who provides guidance and assistance to all facilities throughout the corporation. The next management level is plant management, which is the top management of the individual plant site. The loss control manager should report directly to the plant manager. The internal reporting and external relationships are also indicated. The loss control manager also has internal reporting and external relating relationships. These will vary slightly depending on the size and type of the operation. In addition to the vertical interactions, we must also cultivate horizontal communications. If you have a relatively minor concern about a production operation, you should be able to work this out directly with production management.

Each of the individuals reporting to the plant manager plays an important role in the loss control program. Although none of these individuals is a safety professional and their primary responsibilities are not loss control, their functions are critical. It is important that we sell these people on loss control and encourage them to be actively involved.

The production manager is primarily involved in getting your organization's product, made on time and at a profit. This focus sometimes seems to conflict with loss control efforts, but in reality, it does not. We must understand this and educate

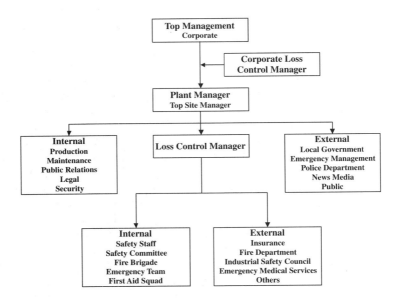

FIGURE 3.5 Organizational structure — large organization, day-to-day.

our co-workers. If a machine is run past the scheduled shutdown for maintenance because "there is no time for a shutdown," the end result may be an unscheduled shutdown caused by a fire on the machine.

If time is invested to educate the production managers, they will become valuable allies in loss control efforts. Production managers will usually cooperate with and contribute to the loss control program once they understand the direct benefits to the daily operations of the plant. Part of running an efficient, profitable operation is effectively handling loss control issues. Production managers and their personnel have a major impact on the success of the loss control program because of their daily involvement with the processes in the facilities. These processes can be the cause of some of the largest potential losses, and it is therefore essential to have the personnel responsible for them on your side and to sell them on loss control.

Maintenance managers are also involved with daily operations on the plant floor similar to production personnel, but they have a different perspective on the operation, which can be important. Maintenance personnel are not directly responsible for producing the product. Their focus is on keeping the process machinery and equipment running smoothly. They also keep the structure itself functioning properly. They are possibly in the best position of all personnel to see problems early enough to prevent a loss. Unfortunately, they are also in a position to create potential losses by applying quick fixes. What action they take depends to a large extent on how well they are trained in and sold on loss control. We must gain their support and cooperation as a crucial element of the overall loss control program.

Public relations department personnel assist with the loss control efforts in two primary areas. They help develop the image that the organization is concerned about loss control. This public image is often underrated. If the organization is perceived as safety conscious, the community views the organization as a better neighbor. Ongoing public relations is important during times when everything is going well, but it becomes essential if a major emergency ever occurs. If you think this public perception is unimportant, remember the headlines concerning Union Carbide's Institute, West Virginia plant or countless other examples of public outcry after an emergency. The public relations department must also be involved in the emergency planning process. How information is handled during an emergency can be very important to the outcome of the emergency. This is particularly true of publicly released information.

Legal departments seem to be present in almost every situation in our litigious society, and that is no less true in loss control. They can help decipher the tangle of regulations, codes, standards, and local ordinances. Occasionally, some aspect of your loss control program may appear to conflict with union agreements, local ordinances, or insurance company requirements. The legal department can help resolve these sometimes difficult issues. We must balance this concern for legal matters and the need to establish a cooperative environment for our loss control program. The involvement of the legal department, particularly in the early stages of discussion, frequently results in a confrontational environment that is counter-productive. Although the legal department may need to be involved in some aspects of the loss control program, discretion should be used regarding both the frequency and scope of their involvement.

The internal security force also plays an important role in the loss control program. The primary loss prevention function of perimeter security is to help avoid losses from arson and sabotage. This threat may become more prevalent in the future. The main benefit of internal security is their constant patrol and observation within the facility. They are the first line of defense in preventing and controlling losses. When properly trained, they can assist in fire prevention and help to ensure that all fire protection equipment is where it should be and is ready for immediate use.

The external relationships that plant management must cultivate are as important as the internal ones. Plant managers must take the lead in developing these relationships. Much of the detail work will, however, be done by those who report to the plant manager. The plant manager should be directly involved in establishing a relationship with local government officials, including emergency management personnel. This relationship must focus on common interests and cooperation for the mutual benefit of the organization and the community as a whole. Open lines of communication are important during day-to-day operations and are critical during emergencies.

The main thrust of the relationship with the local police department will come from the internal security manager. This relationship serves the same purpose as those discussed above, only at a different level. News media and public information contacts are the responsibility of the public relations department. It is important to maintain strong contacts with the local media to help build a good public image. This will be particularly critical during an emergency.

The loss control manager also has internal and external relationships that enable him/her to handle responsibilities effectively. The safety staff, which, depending on the size of the organization, may range from being nonexistent to being very large, assists the loss control manager with the day-to-day safety activities. Even in the smallest organization, it is useful to have at least one part-time assistant to handle situations that may occur when the loss control manager is not at the plant. A safety committee provides a valuable source of ideas and encourages a team effort by all employees. It provides an avenue for direct employee input concerning the loss control program as well as ensures a regular opportunity for employee education. Allowing employees a direct role in the development of the loss control program helps gain their commitment and support.

Fire brigades and emergency teams are an option that will be discussed in detail later. If you decide to use one of these groups in your loss control program, the team will typically report to the loss control manager. The same will usually be true for a first-aid squad.

The loss control manager must maintain a liaison with insurance company representatives. This is especially true for a company insured under a highly protected risk (HPR) concept. The insurance company can frequently provide guidance and resources that will help make a loss control program more effective.

A relationship with the local fire department is important in several areas. The fire department will frequently have resources unavailable to a business and vice versa. Cooperation benefits both organizations. The fire department may provide inspection services that can enhance all loss control efforts. It is essential that the fire department be actively involved in the pre-emergency planning process.

A local industrial safety council is an excellent source of information and ideas. This type of organization usually has monthly meetings which allow for discussions with colleagues, and they frequently provide guest speakers on various topics of current interest.

Local emergency medical services can also be a resource for ongoing operations and are essential when accidents occur. Establishing a relationship with them can be beneficial to both organizations.

There are many other outside resources that can be useful. Companies should get involved with at least one national organization designed for safety and loss control professionals. There are a number of such associations listed in the appendix.

In a smaller organization, most of the organizational structure will be similar but scaled down. Figure 3.6 illustrates one option for organizing loss control in a small company.

EMERGENCY ORGANIZATION

Emergency situations present a challenge to any form of management. The organizational structure during an emergency is illustrated in Figure 3.7 for a large organization and in Figure 3.8 for a small organization. The two main components of the emergency organization are the crisis team (CT) and the emergency operations center (EOC). The purpose of the EOC is to enhance the ability to bring the emergency under control and to stabilize the situation. The purpose of the CT is to manage

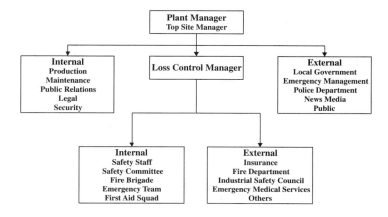

FIGURE 3.6 Organizational structure — small organization, day-to-day.

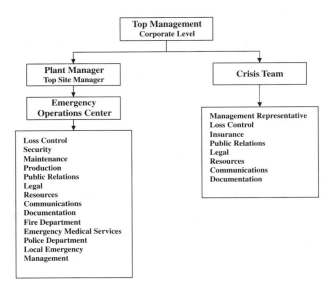

FIGURE 3.7 Organizational structure — large organization, emergency.

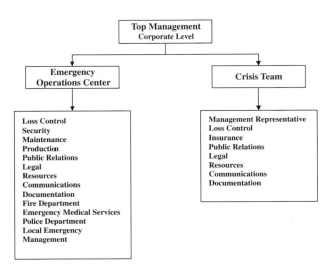

FIGURE 3.8 Organizational structure — small organization, emergency.

the overall impact of the emergency on the organization as a whole. For example, if a major chemical company has a leak that threatens thousands of people, the EOC is primarily concerned with the immediate problems of protecting the population and stopping the leak. The CT, on the other hand, is focused on items like mitigating the damage to the company's image and the impact on continuing operations.

The EOC should be activated anytime there is an emergency at the facility. The extent of the activation should vary depending on the emergency, but the center should always be activated, no matter how small the emergency. The EOC should be used in all cases for several reasons. First, small emergencies have a way of becoming large ones. By activating the EOC at the first sign of trouble, personnel are ready to handle the situation if it gets worse. Also, we tend to react during emergencies the way we have reacted to them in the past. If we do not use the EOC during the more frequent small emergencies, chances are good that habit will make us neglect the EOC if we ever have a major emergency. In addition, practice is needed to operate the EOC effectively during a large emergency. If we do not use every opportunity to activate the EOC, we will never get the practice we need to become effective. Finally, the functions carried out by the EOC are needed regardless of the size of the emergency. Common sense dictates that not all activations will be full scale, but do not fail to activate the EOC to the extent necessary.

The EOC is the control center for the emergency situation. All major decisions regarding the emergency are made here. Communications from the scene of the emergency to all outside organizations are funneled through the EOC. It is the nerve center that directs the operation and ensures a coordinated response to the emergency. The various representatives that comprise the EOC have all been introduced before in the context of their day-to-day relationship to the loss control program. The primary role of the individuals already discussed is to represent their respective organizations at the EOC. Each representative should be the ranking individual on the scene from the organization represented. If the fire chief is on the scene, then he/she should be at the EOC. This allows the maximum cooperation and coordination to apply the various resources at each organization's disposal.

Three new positions are created for emergency operations. These emergency personnel can be from your organization or from one of the others. A company should have its own people assigned to these positions at least for the smaller emergencies that do not require the involvement of outside agencies.

The person in charge of resources is responsible for handling all requests for resources and maintaining an ongoing record of the people on the scene as well as some information about each person. All resources must be monitored, which is especially critical during major emergencies. Keeping a record allows the responsible individual to quickly determine whether the resource is available on scene and, if so, to send it to where it is needed. If the resource is not on scene, he/she locates it and requests that it be brought to the scene.

Communication is essential during an emergency. The person in charge of communications is responsible for all communications from the front line units to the EOC and from the EOC to outside agencies and resources. By channeling all communications through one individual, the process of information exchange is improved.

Documentation is an important and often neglected part of emergency operations. The individual in charge of documentation is responsible for documenting the incident and the actions taken to control it. This effort accomplishes two important tasks. First, it provides a record of the incident that can be used later to help determine ways of improving the response to future emergencies. Second, if legal problems develop, an accurate and detailed record of the incident can be of vital importance.

IMPLEMENTING LOSS CONTROL

Several items must be considered during the implementation of a new program or when making a significant change to an existing program. First, top management must be sold on the idea. To do this, analyze the cost/benefit of the program and be prepared to justify the need. Until adequate justification has been developed, it is pointless to even bring the item to top management for consideration. Prepare for objections in advance. Try to think of all the questions that might arise and how they may be answered. Do your homework concerning any experience with similar programs at other plants within your organization or in outside organizations. How well did the same idea work in those facilities? What can be learned from their experiences? If the program was successful, what was done right that could be duplicated in your facility? If the program failed, what were the reasons, and how do you propose to avoid the problems they experienced. Once you have prepared yourself, prepare your presentation. Anytime you need to make a presentation, the quality and the results of that presentation are directly related to the amount of preparation that went into it. Managers are, as a general rule, very time conscious, so keep the presentation brief and to the point. Tell them what you want to do, why you want to do it, what resources it will require, and the benefits it will provide to the organization. If handouts, visual aids, or other training techniques will improve the presentation, use them.

Each element of the loss control program should contribute directly to the accomplishment of one of the objectives you developed for the program. If this is not the case, you should reevaluate your objectives and programs. Objectives need to be specific, measurable, realistic but not too easy, and written. Objectives set the sights of the program to ensure that all goals are achieved. They cannot provide the necessary direction if they are written in general terms. A method for measuring the progress toward an objective and for alerting you when an objective has been accomplished must be built in. Objectives should challenge you but not be so far out of reach as to seem impossible. This is a delicate balance and will require practice at setting objectives in order to become skilled at achieving this balance. It is important to put the objectives in writing so they are a clear commitment to action and so everyone in the organization is aware of what the objectives are.

In order to gain the support of supervisors and employees, use the same basic technique that you used for management. You have to constantly sell loss control.

Make the benefits to the individuals readily apparent. Ask for the help you need to reach the objectives and make it clear to all concerned that you need their help. This makes everyone feel more a part of the team.

The best way to emphasize safety and loss control in all aspects of operations is to become a highly visible feature on the plant floor. The importance of loss control and its methods needs to be constantly reinforced. When you are on the plant floor, talk with people. Comment on the things they are doing right and encourage them. When you find something that is being done incorrectly, gently correct the problem and remind those involved of the proper procedure. Your constant presence will have a major impact over time and is far more effective than sitting in your office writing memos and procedures. Get directly involved. Your enthusiasm will be contagious.

One of the keys to successfully implementing a loss control program is planning. This should be the first step whenever you introduce a new component or significantly change an existing portion of the loss control program. Planning should answer several key questions: Who will need to be involved in the new program? What will the role of these people be in implementing the program and its ongoing operation? Why are these people necessary? What needs to be done to implement the program? Why is each of these actions important? When should the program be instituted? Why is this timing important? When should each phase of the program take place? Why is this sequence needed? What resources will be needed and when? How will these resources be used? Where will the implementation steps take place? Why is the location important? These questions could go on and on. The ones listed here will give you some ideas, and you can expand your list based on your own individual situation.

Training is the last phase of implementation discussed here. Training is essential to a loss control program. We stated earlier that people present the greatest challenge to effective loss control. Human behavior is at the core of most losses, and training is our primary resource to change the way people behave. During the implementation phase of the loss control program, remember that the program will almost certainly require some type of training.

MANAGING LOSS CONTROL

The continuing success or failure of any loss control program rests with the management of the operation. The personal effectiveness of those managing the loss control program has a major impact on its overall success. We must constantly strive to improve our knowledge of and abilities in technical aspects of loss control, as well as our personal effectiveness. Management personnel must bring enthusiastic leadership to the loss control program. Individual leaders must set an example and reinforce positive performance in others at every opportunity.

In specifically addressing loss control management, one of the first recommendations is to make effective use of your computer. Management deals with information. How effectively you handle the information will often determine the effectiveness of the program. A computer can aid in the management of every phase of loss control. Even if it is only used for record keeping, a computer will easily pay for itself in time savings. The multitude of required records has become almost impossible to maintain without a computer. The computer is also an excellent tool for developing and presenting training programs.

Database software is designed to store and retrieve information of any type. These programs can maintain almost any type of records. Examples include:

- Extinguisher purchase records
- Extinguisher inspection reports
- Installed system inspection and maintenance
- Employee training records
- Supplier lists
- Personal address and phone book
- Safety library list
- Accident reports

An internet connection is another valuable way to use a computer. There is a wealth of information available on the Web concerning loss prevention and control. Also, many regulatory requirements are posted on the Web to allow easy reference.

One specialized type of software that may prove especially useful is a personal information manager (PIM). This type of software will, if used effectively, help you maintain contacts and manage your deadlines.

Another major personal effectiveness issue is time management. If you have not yet developed a workable system for improving the way you managed your own time, you should do so immediately.

4   Life Safety

CHAPTER OBJECTIVES

You will be able to identify and explain:

- General principles concerning life safety
- Human behavior under emergency conditions
- Exits and their essential features and components
- Evacuation plan components
- Principles of smoke and fire control related to life safety

You will also be able to:

- Evaluate the adequacy of exits
- Evaluate overall life safety
- Develop an evacuation plan
- Develop personnel training for life safety
- Conduct an evacuation drill

GENERAL PRINCIPLES

Life safety is the most important reason for a loss control program. The primary goal of life safety efforts is to prevent injury or death. All of the other goals of the loss control program are secondary. Equipment, inventory, and buildings can be replaced; human lives cannot.

People can be protected during an emergency in three ways: (1) by removing the people from the harmful effects of the emergency, (2) by controlling the emergency, and (3) by defending the occupants from the effects of the emergency in place. These options are not mutually exclusive and can be applied in combination. All of the more detailed requirements and practices associated with life safety are designed to contribute to one or more of these basic concepts.

Several general principles concerning life safety exist that can be applied in any situation. Each of these principles involves a crucial area of life safety that must be considered in specific terms for each type of occupancy. The general principles are outlined first to provide a basic understanding of the overall picture regarding life safety. They are not listed in order of priority. The importance of each principle relative to the others is dependent upon the specific situation.

Adequate Number of Exits

Sufficient exits must be available to allow occupants to escape quickly to the outside of the structure. Many major incidents involving a large number of deaths and injuries have occurred over the years. Frequently, the number of fatalities can be at least partially attributed to an inadequate number of exits. How many exits are needed depends on many factors. The type of occupancy, number of occupants, kind of occupants, and type of structure are among the more significant factors. A theater that can hold 1000 people must have more exits than an office designed for fifty.

Exits Clear and Unobstructed

In the event of an emergency, it is essential that occupants be able to use the available exits. Exit doors and exit access areas (the areas leading to the exit) are frequently obstructed or completely blocked. This cannot be permitted under any circumstances. Locking exits is a bad practice and should be avoided. The exits must be maintained in a useable condition. The exit discharge, which is the area immediately outside of the exit door, must also be clear and unobstructed.

Identify Exits and Exit Access

It is of critical importance that the occupants of a structure be able to quickly and readily identify exits and their access routes. Therefore, exits within a facility should be clearly marked as such. In locations where natural lighting provides adequate illumination, unlighted signs are sufficient. When the natural lighting is not sufficient, exit signs must be illuminated. This illumination should be provided with back-up electric power in the event of a loss of normal power. In cases where a hallway, a door, or another structural feature that is not an exit could easily be mistaken for either an exit or access to an exit, it should be clearly marked "Not an Exit."

Adequate Normal and Emergency Lighting

Occupants must be able to locate the access to exits and exit rapidly during an emergency. Adequate normal and emergency lighting is critical because even personnel that are familiar with a facility can become confused and disoriented during the excitement and stress of an emergency. Insufficient lighting will exaggerate this excitement and may make escape impossible.

Provide Time for Escape

During an emergency, time is almost always on the side of the emergency. Conditions in the structure may deteriorate rapidly, making escape impossible. Occupants must be given time to escape, and the two primary methods to increase the time available are alarm systems and control systems. Alarm systems, which can detect fires more quickly than human senses and provide notice of fires in unoccupied areas, offer occupants a significant time advantage. The earlier the fire is detected, the more time will be available before conditions in the structure become dangerous to the occupants. Systems that control the growth and spread of the fire or dangerous fire products (smoke, fire gases, and heat) also increase the time available for escape. The more effectively these methods are used, the greater the opportunity for occupants to escape successfully.

TRAIN ALL PERSONNEL CONCERNING EMERGENCY ACTIONS

Proper responses to emergency situations do not occur by chance. Personnel must receive training concerning the appropriate actions to take in response to an emergency situation. Minimally, individuals should be trained in the procedures to be used when an emergency is discovered, how to escape successfully, and what to do once outside the facility.

DETAILED REQUIREMENTS

The most authoritative source of specific recommendations in the area of life safety is the National Fire Protection Association's (NFPA) Life Safety Code®, which contains detailed recommendations needed for each type of occupancy. The NFPA uses occupancy types including places of assembly, educational, health care, detention and correction, hotel, apartment, lodging or rooming houses, one- and two-family dwellings, mercantile, business, industrial, storage, and unusual structures. In the Life Safety Code®, these occupancy types are also divided into new construction and existing buildings.

General sections of the code specify requirements and concepts that are needed regardless of occupancy. These include means of egress, features of fire protection, building services, fire protection equipment, and definitions.

The Life Safety Code® is referenced extensively by Occupational Safety and Health Administration (OSHA) regulations, and many local jurisdictions have adopted the code as a legal requirement. Anyone responsible for life safety considerations in a facility should have a basic familiarity with the requirements of this code.

HUMAN BEHAVIOR

Successful response to an emergency situation is a combination of prior planning, appropriate occupant behavior, and effective installed systems. Of these three elements, occupant behavior usually has the most significant impact on a specific individual's survival. Planning and installed systems normally have the greatest impact on group survival.

Responding appropriately to an emergency situation is not a natural skill. Many options are available to the individual during an emergency, and people do not always select the most effective option.

Humans, like animals, have an innate survival instinct. Unlike animals, however, people do not have a corresponding set of innate skills. People must be trained to respond appropriately to emergency situations. People are often slow to perceive a threat during an emergency. Once a threat is perceived, the actions taken by an individual can vary considerably. Many options are available; some are effective and others are not.

SUCCESSFUL EMERGENCY RESPONSE PROCESS

Figure 4.1 illustrates the highlights of the process involved in responding to an emergency through the selection of emergency action options. The first step in the process of effectively handling an emergency is notification that an emergency situation exists. This information can come from one or more of three primary sources: alarm systems, verbally from other occupants, or from an individual's senses.

Automatic alarm systems provide the most rapid notification, allowing maximum escape time. Manual alarm systems help speed the notification of other occupants once someone in the structure becomes aware of the emergency. The

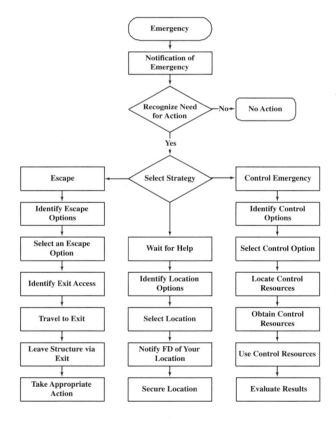

FIGURE 4.1 Emergency action options.

advantage of a manual alarm is the ability to rapidly notify all occupants in all areas of the building that an emergency exists. The major disadvantages of the manual alarm are:

- An individual must discover the emergency first, which delays warning to other occupants.
- The manual nature of the system means that an individual must take action to activate the alarm.
- If the person who discovers the emergency does not think of the alarm or decides that the emergency is not significant enough to activate the alarm, another delay occurs.

Verbal notification of emergencies from one occupant to the next is acceptable only in limited circumstances. Small facilities with only a few employees may be able to make this work effectively. If a shout of "Fire!" cannot be heard throughout the entire facility, this method will not be effective. Detection of emergencies with human senses is the least desirable option. The abilities of occupants to sense the presence of an emergency are considerably slower and more limited than an alarm system's capabilities.

Once notification of an emergency situation has been made by any of the methods just described, the individual must recognize the need for action. This depends mostly on how great the individual perceives the threat to be. If the first impression of the fire alarm signal is "Oh, great, another false alarm," the actions taken by the individual will be slow and not very effective. The opposite end of the perception scale is an "I'm going to die" reaction, which will not bring about the most effective action either. The latter reaction can lead to irrational behavior and taking action without thinking about the consequences. Ideally, occupants will recognize that a real threat is present but will remain calm enough to think through the actions necessary to escape safely.

After an individual has been notified of an existing emergency, he/she must evaluate potential courses of action. The initial decision can be divided into three categories of action:

1. The person can stay where he/she is and wait for help.
2. He/she can exit in a number of directions.
3. The occupant can attempt to control the emergency.

Before the person can select a course of action, he/she must have examined the available options. This is one of the key areas which requires training. To evaluate emergency situations effectively, occupants need information about the impact different circumstances will have on their ability to escape. Normally, an individual on an upper floor of a building would use the elevator to go to the ground floor. During a fire, this could be a disastrous choice. Without the knowledge that this potential escape route carries great risk, the person cannot evaluate it properly.

Once the choices have been evaluated, the individual must select the most appropriate course of action. As soon as the decision to wait, escape, or try to control the emergency has been made, the occupant must identify choices available to him/her in the category that has been selected.

If an escape is to be attempted, he/she must identify the accessible escape routes. Several will usually exist, but not all of them will provide a safe means of escape. Once the options have been analyzed, the individual must select the one that offers the greatest chance of successful escape. To make a successful escape, he/she must identify and reach an exit access, travel through the exit access to the exit, leave the structure via the exit, and take appropriate action once outside the structure.

If the individual decides that the most effective strategy is to wait for help to arrive, he/she must secure the waiting area to the greatest extent possible. This means limiting the potential for the entrance of smoke, fire gases, and heat. The occupant must confirm that someone in a position to help knows that he/she is waiting for help, the exact location, and the current conditions.

If he/she decides to attempt to control the emergency, several items must be evaluated. The fire must be small enough to be easily extinguished. The individual must have access to a means of control, such as a fire extinguisher, and the skills to use it. He/she must consider whether an attempt to control the fire will eliminate other life-saving options. For example, if the fire is close to the only exit from the area, any attempt to control the fire may allow it to grow enough to block the exit. Smoke and fire gases can also become life threatening in a short time.

If all of these decisions have been made carefully, the occupant should be able to select those actions which provide the greatest chance for survival.

EXITS

BASIC CONCEPTS

Exits are a critical part of overall life safety. Several basic concepts are involved in determining the number, arrangement, and construction of exits for a specific area. The number of exits needed is determined by the type of occupancy, the number of occupants, and the hazards present.

Exits include three components:

- Exit access
- Exit
- Exit discharge

The exit access is the area from anywhere in the facility where a person might be leading to the exit. The access in a typical industrial facility may include office, production and storage areas, corridors, and possibly stairs. The exit is the path from inside the structure to the outside; in most industrial settings this includes just the door passing through the exterior wall. The exit discharge is the path from immediately outside the structure to what is referred to in the code as the "public way." This is the nearest point off the property and completely away from the building.

ADEQUATE NUMBER OF EXITS

Sufficient exits must be provided to allow occupants to escape rapidly in the event of an emergency. Only a limited number of people can move through the space of a door in a given period of time. During an emergency, it is necessary to be able to evacuate the occupants of an area in a short period of time. There must be an adequate number of exits to make this possible.

Usually, a minimum of two exits are required from any public area. These exits should be separated to avoid the possibility of a fire quickly and simultaneously blocking both exits. Figure 4.2 illustrates a few examples of exit number and position.

Exit adequacy is calculated using three factors determined by code requirements:

- Exit capacity
- Occupant load
- Travel distance

Exit capacity is determined based upon code requirements. For example, in an ordinary hazard industrial occupancy, 0.2 inches (0.5 cm) of exit capacity is required per person. This means that a minimum size door of 36 inches (91.4 cm) clear width would allow the passage of 180 people.

Width of exit in inches ÷ 0.2 inches (0.5 cm) per person = number of people
36 inches (91.4 cm) ÷ 0.2 inches = 180

Occupant load is the number of square feet of floor area required per occupant. According to NFPA 101®, this is 100 square feet (9.29 square meters) per person in industrial occupancies. This is gross square footage, so nonstructural items within the building do not have to be subtracted from the total.

Travel distance is the maximum distance that must be traveled from any given point in the occupied portion of the structure to reach an exit. For example, in an ordinary hazard industrial occupancy without installed fire protection, the maximum

FIGURE 4.2 Exit number and position.

travel distance to an exit is 200 feet (60.9 meters). Travel distance must be measured along the actual available path of travel, so all obstructions such as equipment, storage racks, and other items must be considered.

When evaluating a building design, all three factors mentioned above must be combined and the factor which permits the fewest people controls the permitted occupancy of the area. For example, the industrial building shown in Figure 4.3 is 60 by 40 feet (18.3 by 12.2 meters) for a total square footage of 2400 (223 square meters). This would permit an occupancy of 24 people. There are two 36-inch (91.4 cm) doors, so the exit capacity of the doors is 360 people. Even assuming that one of the doors was completely blocked, the worst-case travel distance is 100 feet (30.5 meters). In this example, the limiting factor is the square footage. Both the number and size of exits and the travel distance would allow more people to occupy this area.

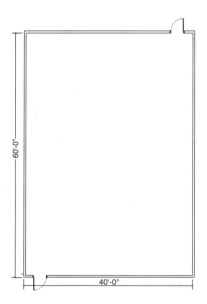

FIGURE 4.3 Life safety example.

ACCESS TO EXITS

Access to exits should be clear and unobstructed. One of the most common problems in the area of life safety is obstructed exits. Policies concerning the obstruction of exits should clearly establish that all exits must be maintained in an unobstructed and usable condition. Figures 4.4 and 4.5 are examples of obstructed exits.

Emergency exits must never be locked when occupants are in the building. If security needs demand that the exits be locked to prevent entry from outside, some type of panic hardware should be used. Figure 4.6 shows a locked exit. Notice the lock on the door at the upper right-hand side of the door. This door could at least be unlatched if the person trying to escape noticed the latch. Figure 4.7 shows an exit with a padlocked security bar above the panic hardware. This arrangement could not be opened by someone trying to escape.

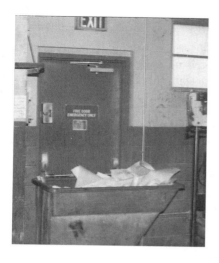

FIGURE 4.4 Obstructed exit.

FIGURE 4.5 Obstructed exit.

FIGURE 4.6 Locked exit.

FIGURE 4.7 Locked exit.

EXIT MARKING

Exits need to be well marked and visible. All exits should be provided with signs indicating that they are exits. If the building area is occupied during darkness or no natural light is available during the day, these exit signs should be illuminated. Figure 4.8 is a photograph of a common illuminated exit sign. The marking of the exit shown in Figure 4.9 is a particularly good example; in that figure, an illuminated exit sign is placed above the exit and a secondary sign is positioned low along the side of the exit. The low sign is used in case smoke might obscure the upper sign during a fire. Figure 4.10 shows an illuminated exit sign for use in hazardous locations.

It should be easy for occupants to determine which potential exit routes are exits and which are not. Any route that could be mistaken for an exit should be clearly marked "Not an Exit" (Figure 4.11).

FIGURE 4.8 Exit sign.

FIGURE 4.9 Exit signs.

FIGURE 4.10 Exit sign for hazardous location.

FIGURE 4.11 Door marked "not an exit."

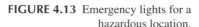

FIGURE 4.12 Emergency lights.

FIGURE 4.13 Emergency lights for a
hazardous location.

EMERGENCY LIGHTING

Emergency lighting is designed to allow occupants to move through the facility and find exits even if the power is off. Emergency lighting is supplied by batteries and should operate for at least 90 minutes after the power has failed. Figure 4.12 is an example of a common emergency light, and Figure 4.13 is an example of emergency lights for a hazardous location.

EVACUATION PLANNING

Evacuation planning is an essential element of any loss control program. The purpose of an evacuation plan is to provide a procedure for the rapid and orderly escape of all occupants of a structure. The plan also provides a method to be used to account for all personnel once out of the structure. OSHA requirements (29 CFR 1910.38) provide adequate guidelines for developing evacuation plans even if the facility in question is not covered by the regulations. OSHA's Emergency Action Plan (EAP) includes the following elements: reporting emergencies, evacuation, and some shutdown requirements.

OSHA regulations demand that several minimum elements be addressed by a plan. Evacuation plans must be in writing unless the facility has ten or fewer employees. They must include emergency escape procedures and routes. Procedures must also be developed for those employees who must remain in the facility to shut down critical equipment and processes. A system must be developed to account for all personnel once the evacuation is complete. If employees have been assigned any duties concerning rescue or first aid, these duties must be specified in the plan. The plan must describe the proper method for reporting fires and other emergencies both within the facility and to outside agencies such as the fire department. The people responsible for the EAP must be specified in the plan.

COMPANY NAME
COMPANY ADDRESS
EMERGENCY ACTION PLAN

Purpose

To establish a plan for the rapid and orderly evacuation of all personnel in the event of an emergency. The safety of all personnel at the site is of paramount importance. This plan is designed to enhance the effectiveness of the response to an emergency.

Scope

This policy applies to all employees, visitors and contractors at the *Company* facility at *Address*.

Evacuation Conditions

Conditions which seriously threaten the safety of personnel and may require the evacuation of the facility may occur within the structure or from a situation outside the structure and include but are not necessarily limited to: fire, explosion, flammable gas leak, toxic gas leak, flammable liquid leak or spill, hazardous material spill or leak, structural collapse, bomb threat, natural disaster, power outage, and severe weather.

What is an Evacuation?

An evacuation requires that all personnel from all areas of the facility leave the building. Personnel will assemble at a designated point outside the facility. An evacuation shall be performed when any condition exists which seriously threatens the safety of all personnel within the facility and/or an event has occurred which requires that all personnel be accounted for.

Authority to Order an Evacuation

The authority to order an evacuation, either complete or partial, rests with *Person-in-charge #1*. In the absence of *Person-in-charge #1*, the authority to order evacuation rests with *Person-in-charge #2*.

In the event that neither of these individuals are present when an evacuation is required, the chain of authority shall follow the normal line of authority which exists during day-to-day operations.

Whoever has the final authority at the time of the emergency is referred to as the person in charge throughout the remainder of this policy.

Evacuation Procedures

Any individual discovering a situation which presents a threat or potential threat to the safety of personnel within the facility shall immediately notify other individuals in the immediate area. The individual will then report the situation to a supervisor.

If it is determined by the supervisor that an evacuation is required, the supervisor will announce the emergency over the paging system. The paging system is accessed by dialing ###.

(continued)

All personnel, upon hearing the emergency announcement over the paging system, will immediately shut down the equipment they are working with and proceed to the nearest exit. Once outside the facility, personnel will go directly to the designated assembly location.

The primary assembly area is *Location #1*. The secondary assembly area is *Location #2*. The attached diagram shows the general layout of the facility. The diagram also indicates the outside assembly location.

Personnel will report to their immediate supervisor at the assembly location immediately after evacuating the building. Once personnel have assembled, it shall be the responsibility of each supervisor to determine that all persons are accounted for and have evacuated the structure. Supervisors will then report to the person in charge. Supervisors will confirm that all their personnel are out of the building or identify anyone that is unaccounted for.

On weekends, nights, holidays, and other times when the plant is not in full operation, all employees are responsible for checking in and out with the watchman on duty. The watchman will be responsible for accounting for all personnel.

If the sprinkler system within the facility is activated, a bell will sound in the area of the main sprinkler system valve. This bell indicates that a sprinkler head is flowing water somewhere in the plant. Employees in the area who hear the sprinkler system bell should respond in the same manner described above for actions they should take if they discover an emergency.

All personnel shall be responsible for visitors whom they are escorting. Contract liaison personnel shall be responsible for any contract employees who are in the facility. When handicapped individuals are in the facility, one person shall be assigned the responsibility for assisting that individual should an evacuation be necessary.

All personnel shall remain at their designated assembly area and await further instructions from their immediate supervisor.

Clearing an Evacuation

No person shall return to an area that has been evacuated until their immediate supervisor has indicated it is safe to do so. No individual shall leave the property during the course of an evacuation unless authority has been given to do so by their immediate supervisor.

Reporting Emergencies

The individual discovering an emergency condition shall immediately notify other employees in the immediate area and their supervisor.

Emergency response organizations such as the fire department, police department, or ambulance shall be notified by telephoning 9 for an outside line then 911. If 911 is not operating properly, call the fire department directly at ###-####.

When emergencies are reported to the outside fire department via 911, the following information should be provided:

(continued)

Nature of the emergency and location [The 911 operator will then transfer
you to a specific dispatcher.]

Business name [*Business name*]

Caller's name

Address [*Address*]

Nature of the emergency [fire, explosion, etc.]

Extent of the emergency

Injuries, if any, to personnel

Exact location within the facility

Point of contact meeting place

Call back number [*Call back number*]

Prepared *Date*

Person responsible for this plan *Name and Job Title.*

Evacuation diagrams can be designed in several basic formats. The most effective diagram style will depend on the specific facility, but all diagrams should at least
show the layout of the facility, exits, and assembly points. Figure 4.14 illustrates the
most basic type of evacuation diagram. It shows only the exits from the building and
the assembly points. There is only one primary assembly point and a backup location if the primary cannot be used. This concept is effective for facilities with a small
number of employees.

Figure 4.15 depicts an evacuation diagram which shows exits and assembly
points and divides the building into evacuation areas. In this example, any occupants
in area A would go to assembly area A. Once the employees are accounted for at each
assembly area, a report is sent to the control point. This system works well in larger
facilities with too many employees to effectively assemble at one point.

Figure 4.16 illustrates an evacuation diagram which uses a "You are Here" indication at each location where the diagram is posted. This added complexity is often
unnecessary, but in some situations it is beneficial. If, due to the size of the facility,
employees may not be familiar with all areas of the building, it can be helpful. Each
diagram must be prepared specifically for the location where it will be posted. This
increases the time and effort needed to prepare the diagrams. The most common example of this style is the diagram in a motel room. This is the most effective type of
diagram when the occupants are not familiar with the building.

EMERGENCY ACTION PLAN EXERCISES

Exercises are a critical part of the process of planning emergency actions. A plan that
has been developed, however carefully, but not tested through the use of an exercise
is not proven. Even a plan that has been tested may break down under actual emergency conditions, but it is less likely to prove inadequate if it has been tested. It
should never be assumed that a plan will work without testing. Planning for emergency situations is difficult at best, and exercises provide a reality check that is
needed to ensure that what looks practical on paper also works in practice.

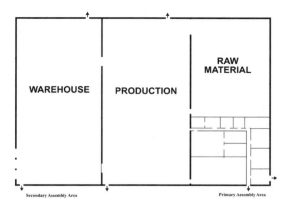

FIGURE 4.14 Evacuation diagram.

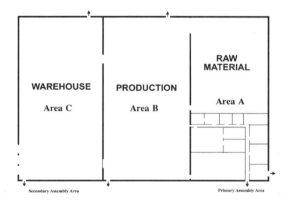

FIGURE 4.15 Evacuation diagram.

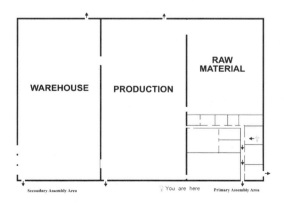

FIGURE 4.16 Evacuation diagram.

All planning elements should be completed prior to conducting an exercise. Employees should be briefed on the overall plan and their specific roles and responsibilities under the plan.

The first exercise for all but the simplest plans should be a walk-through. This type of exercise is conducted to familiarize all personnel with the process and to identify problems in the plan. Actions are evaluated for accuracy and results, not speed. Emphasis must be placed on the performance of the plan. Criticism of individual actions should be avoided.

Planning for the exercise begins with the development of a realistic scenario. What type of emergency will be simulated? Who will activate the alarm? How will initial notification be made? Many other elements can be added to the scenario depending on the specific situation. Always develop a signal to be used in the event a real emergency occurs during the exercise.

Observers should be selected for use during the exercise. Sufficient observers should be available to monitor the progress of the evacuation at all critical points inside the structure and at all personnel assembly and control areas outside the facility. These observers may be employees or outside personnel. It may be possible to get assistance from your local fire department, a local safety association, and many others.

The date and exact time of the exercise should be determined at least two weeks in advance. The fire department, insurance company, and alarm company should be notified of the exercise. For the first exercise, the date should be posted on organization bulletin boards, and employees should be told of the exercise by their supervisors at least one week prior to it. Personnel should understand that the purpose of the exercise is to evaluate the emergency action plan and to allow them to become familiar with it.

Observers should be briefed on their responsibilities on the day of the exercise. The evaluation should focus on the plan, and each observer should be prepared to give a report on his/her evaluation after the exercise.

Get final confirmation that the exercise can proceed as planned on the scheduled day. A short time prior to the exercise, position observers in the assigned locations. Call the alarm company and remind them that you are going to be activating the alarm. If the alarm signal is transmitted automatically to the fire department, call them.

Start the scenario by activating the alarm or by any other means that has been planned. Monitor the progress of the planned actions inside and outside the structure. Confirm that all personnel have been accounted for outside the building. Make brief remarks at each assembly point expressing thanks for participation in the exercise and point out something that went well and according to plan. Release the employees to go back to work.

All the observers should meet immediately after the exercise to discuss how the actual exercise compared with the actions called for in the plan. Suggestions for improvement in the plan should be developed immediately if problems are obvious. A written report on the exercise, including any recommended changes in the emergency action plan, should be prepared. All supervisors should be informed of the results of the exercise and be directed to brief their employees. Positive comments concerning the exercise should be posted on the bulletin board.

If only minor problems occur during the exercise, another walk-through exercise is probably not necessary. If there were many problems or very significant ones, a second walk-through exercise should be conducted after the plan is revised. One or two months after a successful walk-through exercise, a surprise drill should be conducted. The process for this type of exercise is essentially the same, except that only a few key personnel in the organization know about the drill in advance.

SMOKE AND FIRE CONTROL

One of the key elements of life safety is effective control of the spread of smoke and fire. To provide occupants with the maximum time possible for escape, the hazardous products of the fire must be contained.

Structures should have built-in features to limit the spread of smoke and fire. These features can be active or passive. Active systems perform a task that contributes to smoke or fire control, while passive ones function without any action by nature of the design. All of these features and systems are intended to increase the time available for occupants to escape or make escape unnecessary.

All installed extinguishing systems are active systems that control or extinguish a fire. Smoke control doors are another active system. These doors detect a smoke condition and close automatically. Dampers within the duct work of a ventilation system are designed to reduce the spread of smoke. Automatic roof vents provide ventilation which can help control the spread and development of a fire. Fire doors can help reduce fire spread.

Passive features are components of the structure which aid in smoke and fire control. For example, fire walls are used to separate a building into fire areas. These divisions are designed to limit the spread of fire. Draft curtains, solid barriers suspended below the ceiling, are intended to confine the initial spread of smoke and heat. Many other systems and features are available and will be covered in greater detail in the chapter on installed systems.

PERSONNEL TRAINING

All personnel must receive basic training in how to respond properly to an emergency situation. It is often assumed that people have the necessary skills to respond adequately to an emergency situation without training. This is a false assumption. Personnel must be trained to take appropriate action during an emergency, and this training must cover certain basic elements. Personnel should know how to report emergencies both within the facility and to outside organizations such as fire departments or ambulance services.

Delay in reporting emergencies to the fire department has been a contributing factor in many large loss fires. Occupants need time to escape; delaying an alarm reduces the time available and can lead directly to deaths or injuries.

Personnel must be trained in the appropriate evacuation procedures. A rapid and orderly evacuation will not occur without this training. If personnel are allowed to determine their own courses of action, the evacuation will degenerate into a disorganized and ineffective mass of people. If training has not informed personnel where to go during an evacuation, it will be impossible to accurately account for all occupants.

LIFE SAFETY EVALUATION

All aspects of your life safety program should be evaluated. Most of the following items should be checked at least monthly. Several items concern the facility as a whole and should be covered as design issues. They are included in the evaluation as a reminder to look for changes within the facility that may have an impact on life safety. This list includes the following:

- Two ways out from all areas
- No exits through hazardous areas
- Aisles clear and unobstructed
- Evacuation alarm audible in all areas
- Public address system audible in all areas
- Emergency phone number posted on all phones
- Evacuation diagrams posted
- Outside assembly areas clear and unobstructed

There are also issues relative to each exit that should be considered. These include:

- Sign posted and in good condition (legible, light working if illuminated type sign)
- Access clear and unobstructed
- Emergency lighting present, correctly aimed, and working
- Unlocked doors
- Discharge clear and unobstructed

5 Hazard Control

CHAPTER OBJECTIVES

You will be able to identify and explain:

- Principles of hazard control
- Types of hazards
- Human factors in hazard control

You will also be able to:

- Conduct a hazard analysis
- Develop a fire prevention plan
- Develop hazard control policies

HAZARD CONTROL VS. FIRE PREVENTION

Most of the material discussed in this chapter has traditionally been called fire prevention. The term fire prevention, however, implies that the activities are only successful if a fire is avoided completely. This is not the case. The term hazard control more accurately describes what actually occurs. The activities which comprise hazard control do attempt to prevent fires, but even if a fire is not completely avoided, a great deal is still accomplished if lives are saved and damage minimized.

HAZARD CONTROL HIERARCHY

The hazard control hierarchy consists of six major approaches to handling hazards.

- Eliminate
- Substitute
- Isolate
- Engineering controls
- Administrative controls
- Personal protective equipment

This approach should be applied from the top down. Ideally, all hazards should be completely eliminated. If elimination is not possible or practical, and frequently it will not be, then substitution should be considered, and so forth proceeding down the list.

Elimination is the complete removal of the hazard. The complete elimination of a hazard is often not possible, but when it can be accomplished it has a major impact on the safety of the operation. If a process normally requires a flammable solvent and a way to perform the same work with any solvent is developed, the inherent hazards of the process are greatly reduced.

Substitution is when a more hazardous component of a system is traded for a less hazardous one. The printing industry offers an excellent example of this concept in practice. Years ago, solvents including toluene, xylene, and methyl ethyl ketone were common throughout the industry. These solvents pose serious fire risks and also personnel exposure issues. The printing industry has substituted less hazardous solvents such as isopropyl alcohol in many operations. Alcohol is also flammable but is not as high a risk as the previously mentioned solvents.

Isolation involves placing the hazard in a remote or contained area away from other portions of the operation. The explosives manufacturing industry has done this practically from the beginning of their industry. Production areas are separated by distance and/or blast walls so that an explosion in one area is unlikely to affect the entire facility.

Engineering controls are systems that should not require direct human intervention to work effectively. These may be used in both prevention and control areas. A prevention example is a spring-loaded valve handle on a flammable liquid dispensing arrangement. This is designed to prevent an individual from opening the valve and leaving the area with flammable liquid still flowing.

These controls are systems that monitor conditions within a process and can shut down the process automatically if some aspect of the process gets out of an acceptable range. Temperature limit switches are an example of this type of control. A furnace used for heat-treating metal parts will be equipped with a high temperature limit switch separate from the thermostat control. This limit switch will shut down the system if temperatures exceed a predetermined upper limit. Fire protection systems are also engineered controls. These systems are designed to operate automatically and control or extinguish a fire.

Engineering controls are only as effective as people make and maintain them. An improperly designed control will not function correctly when needed. All engineering controls will require some level of maintenance and occasional repair. If a system is not maintained properly, it may not perform as designed when needed. Engineering controls may also be intentionally defeated. The spring-loaded valve mentioned above may be propped open by an operator to allow him/her to do something else while a container fills, thus defeating the purpose of the valve.

Administrative controls include policies, procedures, rules, and training. Administrative controls are less effective than the previous elements of the hierarchy because they rely heavily on people for effective implementation and consistent use. They are, however, still a very important part of an effective approach to hazard control. There are many occasions where engineering controls must be supported by administrative controls. For example, bonding and grounding of flammable liquid containers during transfer operations is an engineering control. Administrative controls including procedures for use and appropriate training must be in place for this system to function properly.

Personal protective equipment (PPE) is the last line of defense for hazards that pose a threat to people. For example, fire retardant clothing is a common requirement in petroleum refineries. Much effort is expended to ensure that fires do not occur, but the nature of the materials and processes makes it likely that at some point there might be a fire. Fire retardant clothing provides an extra measure of protection to the people that work in these areas.

GENERAL PRINCIPLES OF HAZARD CONTROL

Hazard control is the process of using resources to eliminate or reduce the opportunities for fires to start and minimize their impact when they do occur. As was previously discussed, the fire triangle is a concept which states that sufficient heat, fuel, and oxygen must be present for a fire to exist. In practice, hazard control usually requires a primary focus on one of these elements, usually the fuel or heat. A welding shop, for example, must have heat. The heat used to weld and cut metals cannot be eliminated. In this case, hazard control efforts will focus primarily on controlling combustibles. The opposite is true in an office. There must be combustibles or the functions of the office cannot be maintained. The emphasis in that case is on controlling ignition sources. Both ignition sources and ignitable materials are considered, but the primary emphasis is usually on only one of these.

Separation of operations and hazards is a basic concept in loss control. The principle assumes that operations that are separated are less likely to be lost in the same fire. Separation accomplishes two things. First, it reduces the potential for a fire in the more hazardous area to spread to other areas. Second, it reduces the probability that an ignition source in the low-hazard area will cause a fire in the high-hazard area. The separation of portions of a process is most commonly used for high-hazard processes. For example, the hazards of a paint spray booth that is part of a manufacturing process can be reduced by separating the spray booth from other operations. This separation may involve fire walls, distance, or other methods. An extreme example of failure to separate properly would be placement of a welding shop adjacent to a flammable liquid storage room. Separation can also be used to divide large areas even if no high-hazard process exists.

TYPES OF HAZARDS

In relation to fire protection, hazards produce or contribute to one or more of three factors: the ignition of a fire, a material which can be ignited, or the growth and spread of the fire. Hazards may have more than one of these basic characteristics.

SPECIFIC HAZARDS

Smoking

Smoking materials provide an ignition source, and careless smoking practices are responsible for many fires. The smoking hazard can be controlled in one of two ways.

First, the smoker can be restrained with policies that specify where and under what circumstances smoking is permitted. This approach depends on people being educated about the policy and then complying with it. In most cases this is sufficient. The second method is the total elimination of smoking by prohibiting smoking materials in an area. This approach requires personnel to deposit all smoking materials outside the area prior to entering. This strict method is usually reserved for extremely hazardous areas such as chemical refining and explosive-manufacturing operations. It eliminates the possibility of an individual forgetting or ignoring the no-smoking policy.

Regardless of your opinion about smoking, those who are smokers will smoke. Unless there is an essential fire prevention reason for the complete prohibition of smoking, a designated area for smoking where fire prevention issues can be addressed effectively should be provided. If no area is designated, smokers will have a tendency to find hidden places in which to smoke, and this can create a much more hazardous situation.

In areas where smoking is not permitted, the no-smoking policy should be posted conspicuously and be enforced consistently. Ashtrays should be made available at all points of entry from areas where smoking is permitted. Wherever smoking is permitted, an adequate number of ashtrays and disposal facilities should be available. Figure 5.1 is an example of a common smoking material disposal device.

Housekeeping

Housekeeping refers to the general order and cleanliness of the operation. This is one of the key areas in assessing a facility's commitment to loss control. A facility which is clean and orderly in appearance usually places more emphasis on loss prevention

FIGURE 5.1 Smoking material disposal container. (Photo courtesy of Justrite Manufacturing Company.)

and control than a facility that is disorganized and messy. Housekeeping addresses basic issues that are easily overlooked or placed in low priority status. They are simple items that often appear insignificant. Ignoring them, however, can lead to larger problems.

Waste control is one of the basics of housekeeping. Waste materials should always be accumulated in appropriate containers. Trash receptacles must be emptied regularly. This is usually necessary at least once a day. One type of trash receptacle that is intended to be self-extinguishing is illustrated in Figures 5.2 and 5.3. The smoke and other products of combustion are directed back into the container, thus eliminating the oxygen around the fire, causing it to go out. Waste materials should not be allowed to accumulate within the facility. Contaminated waste materials, especially those containing flammable or combustible liquids, should be discarded in metal containers with self-closing lids (Figure 5.4).

Failure to control the accumulation of combustible materials (Figure 5.5) provides an opportunity for fire to occur.

Equipment should be as clean and free of grease and dirt as is practical. Combustible scrap should not be permitted to collect around equipment. Fluid leaks,

FIGURE 5.2 Trash container. (Photo courtesy of Justrite Manufacturing Company.)

FIGURE 5.3 Trash container cutaway. (Photo courtesy of Justrite Manufacturing Company.)

FIGURE 5.4 Rag container. (Photo courtesy of Justrite Manufacturing Company.)

FIGURE 5.5 Accumulation of combustible material.

particularly of flammable or combustible liquids, should be corrected and cleaned up immediately. Leaks which are an inherent part of the operation, such as small hydraulic fluid leaks, should be absorbed with an appropriate material and cleaned up regularly.

Packaging materials should be stored in noncombustible bins with self-closing lids. The amount of packaging materials maintained in the work area should be kept to a minimum — usually one day's supply. Locations where packaging is done should be designated as no-smoking areas.

Many industrial operations create dust. The control of dust is also an element of housekeeping. Dust provides an excellent fuel because the finely divided particles are easily ignited and burn rapidly. Dust accumulates on most surfaces and provides an easy way for fire to spread. Dust may explode under certain circumstances. This can be a particularly dangerous problem during cleaning operations if dust is blown down using compressed air. The dust clouds thereby created can be extremely hazardous. Generally, dust should be vacuumed, not blown.

The most effective method of handling dust is to prevent its accumulation with effective dust collection systems. All sources of significant quantities of dust should be ventilated into a dust collection system. This method avoids the hazards of both the dust accumulation and the associated cleanup operations.

Outdoor areas should be maintained as free of combustibles as possible. Do not collect combustible materials, particularly in locations close to the building. Items such as pallets (Figure 5.6), combustible inventory or raw materials, and dumpsters should be separated from the building. Brush and grass should be cut short close to the building. Fire lanes at least 20 feet wide (6 meters) and 10 feet (3 meters) from the building should be maintained on all paved areas.

In storage areas, aisle widths and clearances are two of the most significant concerns. Aisle spaces can retard the spread of fire. They act as a fire break because there is no fuel in the aisle, and it is difficult for the fire to transfer across it to the next rack. In addition, aisles also provide access for manual fire fighting. Clearances are another consideration in storage areas. A two-foot space should be maintained between stored materials and walls. This space allows materials that are water-absorbing to expand and provides access for fire fighting. A 1-foot (30.5 cm) clearance from ceilings should be preserved. If the area has sprinklers, stock should be placed no closer than 18 inches (45.7 cm) below the sprinklers.

FIGURE 5.6 Pallets.

Even combustible materials that are properly stored need to be considered. Fire loading is the evaluation of the fuel contribution the materials in an area would make to a fire. Areas with items that contribute a significant fuel value need special emphasis, for example, roll paper storage (Figure 5.7), foam storage (Figure 5.8), baled paper waste (Figure 5.9), and tire storage (Figure 5.10).

FIGURE 5.7 Roll paper storage.

FIGURE 5.8 Foam roll storage.

FIGURE 5.9 Baled waste paper storage.

FIGURE 5.10 Tire storage.

Heat-Producing Equipment

Heat-producing equipment provides a source of ignition. Heat-producing equipment in industry is primarily used for either comfort control or process equipment. Heat may be produced by burning a number of different types of fuel including various petroleum products such as gasoline, kerosene, and fuel oil or flammable gases such as natural gas (methane) and propane. Electricity may also be used to produce heat.

Building and area heating equipment includes central systems, fixed area heaters, and portable devices. Process heat-producing equipment includes everything from soldering irons to electric arc furnaces for melting large quantities of metal. Each of these processes has unique issues which must be addressed. Space limitations restrict present coverage to general guidelines.

The two main areas which should be addressed to reduce the potential for igniting a fire are equipment maintenance and proper clearances. All devices should be Underwriters' Laboratories (UL) approved. This approval establishes that the equipment has met certain basic safety criteria. Regular cleaning and preventive maintenance should be conducted, and the manufacturer's recommendations concerning maintenance should be followed. The entire fuel chain should be inspected regularly. Fuel chain refers to the path of fuel from storage until it is consumed by a device. Fuel chain inspections should discover leaks of flammable or combustible materials. The wiring of electrical equipment should also be checked periodically for wear and damage.

Clearances are important due to the heat generated by these devices. The appropriate clearance between the equipment and any combustible material will vary depending on the type of device and the amount of heat produced. Follow the equipment manufacturer's recommendations. After the equipment is installed, ensure that clearances are maintained through regular inspections.

Portable space heaters, because they are not mounted permanently, can cause special clearance problems. Since the user must determine the appropriate clearance from combustible materials, individuals that are permitted to use space heaters need training in the proper use of these devices.

Electrical

Electricity is a common cause of fires. Its widespread and varied use in most facilities requires special attention during hazard control. As mentioned above, all equipment should be UL approved. This ensures that the device itself is fundamentally sound from an electrical safety perspective.

All electrical installations should be made by qualified personnel and according to the electrical code in force locally. Circuit overloading must be avoided. When electrical circuits draw more current than the wiring was designed for, the wires can overheat and cause a fire. Fuses and circuit breakers of appropriate amp ratings should be used to prevent circuit overloading. The condition of electrical cords and cables should be checked regularly. Worn insulation and broken conductors can expose combustibles to electrical current or arcs that can cause ignition. Extension cords should not be used as a substitute for fixed wiring because they increase the risk of circuit overloading and exposed electrical conductors. Electrical equipment should receive periodic preventive maintenance to ensure that problems are discovered and corrected before they can cause arcs or shorts that may ignite a fire.

Flammable and Combustible Liquids

Flammable and combustible liquids are used for fuels, solvents, lubricants, and many other purposes in most facilities. Some types of operations require more than others, but few operations are without at least some of these materials.

To help determine necessary hazard control measures, flammable and combustible liquids are classified according to their degree of hazard. The degree of hazard is based upon the flashpoint and boiling point of the liquid. Hazard control measures must be more stringent for the more hazardous liquids. The table below illustrates these classifications. The classification system is used primarily to determine the appropriate code requirements for storage, transportation, use, and protection. For example, less class IA liquids would be permitted in a storage area than class IC liquids due to the greater risk.

Storage

Flammable and combustible liquid storage should be designed to prevent ignitable vapors from reaching an ignition source. Loss prevention should address both of these issues — controlling the flammable liquid and its vapors and ignition sources. These control issues must be maintained throughout the storage, dispensing, use, and disposal phases of the flammable liquids process.

The flammable control is accomplished in two phases. The actual liquid container is the first phase, and the storage area for the containers is the second. Liquid containers made for five gallons or less are referred to as safety cans. Safety cans are

FLAMMABLE LIQUIDS

A flammable liquid has a flashpoint below 100°F and a vapor pressure not exceeding 40 pounds/square inch (absolute) at 100°F.

CLASS IA	Flammable liquids with a flashpoint below 73°F and boiling point below 100°F.	Examples: Acetaldehyde, Ethylene Oxide, Ethyl Ether, Methyl Chloride, Methyl Ethyl Ketone.
CLASS IB	Flammable liquids with a flashpoint below 73°F and boiling point at or above 100°F.	Examples: Acetone, Ethyl Alcohol, Gasoline, Hexane, Methanol, Toluene.
CLASS IC	Flammable liquids with a flashpoint at or above 73°F and below 100°F.	Examples: Butyl Acetate, Butyl Alcohol, Propyl Alcohol, Xylene.

COMBUSTIBLE LIQUIDS

A combustible liquid has a flashpoint at or above 100°F.

CLASS II	Combustible liquids with a flashpoint at or above 100°F and below 140°F.
CLASS IIIA	Combustible liquids with a flashpoint at or above 140°F and below 200°F.
CLASS IIIB	Combustible liquids with a flashpoint at or above 200°F.

Source: Flammable and Combustible Liquids Code, National Fire Protection Association, Quincy, MA.
Note: 73°F = 22.80°C, 100°F = 37.7°C, 140°F = 60°C, 200°F = 93.3°C, 40 psi = 276 kilopascals.

designed to control flammable vapors and contain the liquid. They provide a means for safely transporting and storing flammable liquids. All safety cans must be approved for use by a recognized testing organization such as Underwriters' Laboratories or Factory Mutual. Safety cans must possess certain basic features to perform effectively. They must be leak-proof, automatically vent excessive pressures to avoid container rupture, prevent external flames from coming into contact with the internal flammable liquid, and have automatically closing fill and dispense openings. All containers for flammable liquids should be clearly marked indicating their contents. Safety cans are available in two styles: type I (Figure 5.11) and type II (Figure 5.12). Type I safety cans are filled through and dispensed from a single opening. These cans have a large discharge opening and are primarily designed for dispensing liquids into large containers. Funnel attachments are available for type I cans that allow dispensing into narrow openings. Type II safety cans have separate fill and dispense openings as well as an attached, flexible dispensing spout. Safety cans that are constructed from polymer materials (Figure 5.13) are also available.

Safety cans and other small containers should be stored in flammable liquid cabinets (Figures 5.14 and 5.15) when not in use. These cabinets are generally yellow and must be labeled, usually in red letters, "Flammable — Keep Fire Away." The amount

FIGURE 5.11 Safety can, type I. (Photo courtesy of Justrite Manufacturing Company.)

FIGURE 5.12 Safety can, type II. (Photo courtesy of Justrite Manufacturing Company.)

FIGURE 5.13 Safety can, poly. (Photo courtesy of Justrite Manufacturing Company.)

of a liquid that can be stored in a locker is based on the degree of hazard the liquid presents. The number of cabinets in any single area is also restricted by fire codes.

Cabinets are typically constructed of metal with an air space between the exterior and interior sheeting. They are equipped with a catch basin in the bottom designed to contain a limited amount of spilled material. They are also equipped with a bung opening that may either be used for venting the cabinet or capped. If the cabinet is vented, the vent must be to the outside of the building. The bung opening with a cap in place is shown in Figure 5.16.

FIGURE 5.14 Flammable liquid storage cabinet. (Photo courtesy of Justrite
Manufacturing Company.)

FIGURE 5.15 Flammable liquid storage cabinet. (Photo courtesy of
Justrite Manufacturing Company.)

FIGURE 5.16 Flammable liquid storage
cabinet bung.

Drum cabinets should be used to enclose drums of flammable liquids that must be stored inside the structure for dispensing (Figures 5.17 and 5.18). Drums for waste accumulation should also be in cabinets (Figure 5.19).

Larger quantities of flammable and combustible liquids should be stored in remote areas away from process operations. The ideal storage arrangement is a separate shed (Figure 5.20) or building (Figure 5.21). This provides the greatest safety factor because the separation of the building prevents vapors or spilled material in the storage area from coming into contact with ignition sources in the facility's main

FIGURE 5.17 Drum cabinet, horizontal. (Photo courtesy of Justrite Manufacturing Company.)

FIGURE 5.18 Drum cabinet, vertical. (Photo courtesy of Justrite Manufacturing Company.)

FIGURE 5.19 Drum cabinet, waste accumulation. (Photo courtesy of Justrite Manufacturing Company.)

FIGURE 5.20 Outside storage shed. (Photo courtesy of Justrite Manufacturing Company.)

building. If a fire occurs in the separated flammable liquids storage area, it is not likely to spread to the main building areas. The second-choice arrangement is an attached building with an exterior entrance (Figure 5.22). This room is physically attached to the building but does not have an opening directly into the main building. The third option is an attached building connected by an interior door (Figure 5.23). A fourth method is a separate room within the building that has exterior access (Figure 5.24). This arrangement allows the interior walls to be constructed without the penetration of a doorway and improves fire containment. The final common arrangement is an interior room with interior access (Figure 5.25). The requirements of operations may make an arrangement with interior access the only practical choice. If this is the case, vigilance will be required to ensure that the openings into the room are kept closed. These storage rooms may often be equipped with a fork-lift-size opening in addition to personnel doors to allow for the transport of pallets of material.

All storage areas should have several basic features. Every door should be a rated self-closing fire door, and all doorway openings should be provided with curbing to prevent the escape of liquid. The areas should be equipped with suitable drainage to trap liquids in a containment sump or underground tank. The drains, however, should not be connected to regular sewer lines. Floor level ventilation (Figure 5.26) should be provided to remove flammable vapors, and explosion-proof electrical equipment should be used.

Explosion-proof electrical equipment is designed to avoid electrical ignition of the flammable vapors and to contain any vapor explosion that might occur inside the electrical device. Examples of explosion-proof fixtures include a light (Figures 5.27 and 5.28), junction box (Figure 5.29), control switch box (Figure 5.30), outlet (Figure 5.31), and circuit breaker panel (Figure 5.32).

In addition, fire detection and suppression systems of the appropriate type for the specific hazard should be installed. Installed systems are discussed in Chapter 7.

Dispensing
One of the most common dispensing arrangements is transferring flammable liquids from drums into smaller containers. A typical arrangement of this type is shown in Figure 5.33. Drums should be fitted with self-closing faucets (Figure 5.34) and pressure relief vents (Figure 5.35). A drip can is used to prevent spills incidental to dispensing. The drip pan is designed to contain small amounts of flammable liquids that may drip from storage containers.

Static electricity is a significant ignition risk when transferring flammable liquids. During dispensing and any other product transfer operation, containers must be grounded and bonded to prevent the development of a difference in electrical potential that might result in an electric arc. Grounding is accomplished by establishing an electrically conductive path between a container and an electrical ground. Bonding is achieved by making an electrically conductive connection between two or more containers. Both the dispensing and receiving containers should be grounded and the two containers should be bonded together. Typical clamps used for grounding and bonding are shown in Figure 5.36.

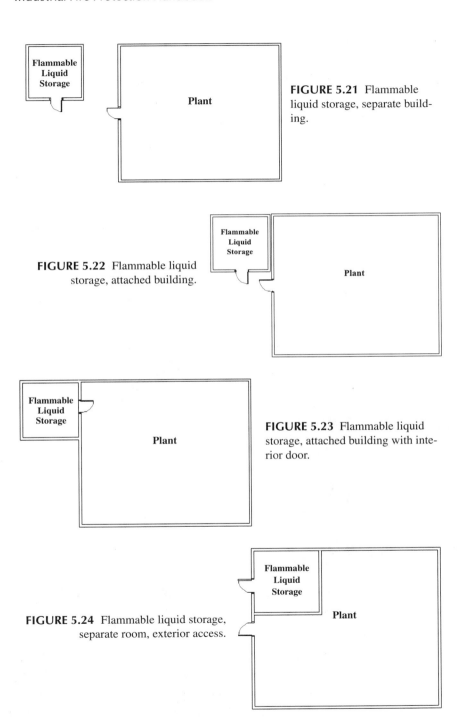

FIGURE 5.21 Flammable liquid storage, separate building.

FIGURE 5.22 Flammable liquid storage, attached building.

FIGURE 5.23 Flammable liquid storage, attached building with interior door.

FIGURE 5.24 Flammable liquid storage, separate room, exterior access.

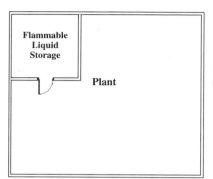

FIGURE 5.25 Flammable liquid storage, separate room, interior access.

FIGURE 5.26 Floor level ventilation.

FIGURE 5.27 Explosion-proof light, incandescent.

FIGURE 5.28
Explosion-proof light,
fluorescent.

FIGURE 5.29 Explosion-proof junction box.

FIGURE 5.30 Explosion-proof control box.

FIGURE 5.31 Explosion-proof outlet.

FIGURE 5.32 Explosion-proof circuit breaker panel.

FIGURE 5.33 Drum dispensing arrangement. (Photo courtesy of Justrite Manufacturing Company.)

FIGURE 5.34 Self-closing faucet. (Photo courtesy of Justrite Manufacturing Company.)

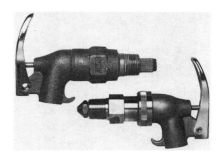

FIGURE 5.35 Pressure relief vents. (Photo courtesy of Justrite Manufacturing Company.)

When many small containers of flammable liquids are used throughout the work area, a portable solvent tank may be used to store, transport, and dispense the flammable liquid. A portable solvent tank is a large tank equipped with a pump that may be wheeled through the facility (Figure 5.37). This allows the flammable liquid to be safely transported to the area of use, and it is more efficient than taking each small container to a central storage area.

Use
The use of flammable liquids always involves a certain degree of risk. However, fires can be prevented through the employment of procedures, safeguards, and vigilance. All personnel involved in the use of flammable and combustible liquids should be thoroughly trained in safe operating procedures and emergency actions.

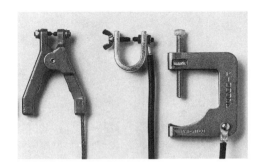

FIGURE 5.36 Grounding cable clamps. (Photo courtesy of Justrite Manufacturing Company.)

FIGURE 5.37 Flammable liquid dispensing cart. (Photo courtesy of Justrite Manufacturing Company.)

Generally, the methods for controlling the hazards of flammable liquids are containment of the liquids, restrictions on the quantities of liquids in the building, safe work practices, adequate personnel training, and installed fire protection systems. Each specific risk should be evaluated using these basic criteria. An example of effective containment is the use of a plunger can (Figure 5.38) in situations where a solvent must be dispensed onto a cloth for cleaning materials.

The following sections provide guidance on some of the risks involved with several common uses of flammable liquids.

Spray Operations

Spray application of flammable liquids creates several risks which can be minimized by using proper equipment and procedures. Spray operations should be separated from other processes in order to prevent ignition sources in less hazardous areas from igniting a fire in the spray operation. Separation also reduces the possibility that a fire in the spray area will spread to surrounding areas. Spray operations should not be located below grade because below grade areas may allow flammable vapors to accumulate. Exits from the spray area must be readily available, and no exits from other operations should have to pass through the spray area.

Ventilation is important to remove flammable vapors. The ventilation system should provide ample air flow to prevent the accumulation of flammable vapors. Ventilation from spray operations should discharge directly outside the structure. In addition, ducts should have adequate access doors and panels for regular cleaning. The ventilation system and all filters should be cleaned frequently enough to prevent the accumulation of flammable residue. Automatic spray operations should be linked

FIGURE 5.38 Plunger cans. (Photo courtesy of Justrite Manufacturing Company.)

with the ventilation system so that the process shuts down automatically if the ventilation system fails.

Installed fire protection systems should be provided in the spray area and the duct work of the ventilation system. In addition to the specific system for the spray operation, sprinklers should be provided. The area should be designated "no smoking." General housekeeping measures should be used to avoid the accumulation of waste materials, and general storage should be prohibited.

All piping for the spray operation should be clearly identified. Valves should be readily available and marked to indicate the portion of the process they control. On manual spray operations, a valve should be placed where the fixed piping terminates into the flexible hose.

Operator training is critical to the safe operation of spray processes. Personnel should be familiar with the materials used and the hazards associated with each material. They should know the location of flammable liquid piping and valves. Operating personnel need to be thoroughly familiar with both normal and emergency operating procedures.

Dipping Operations

The basic protection concepts used for spray operations also apply to dipping operations. One additional required feature is overflow protection for the dipping tank. To reduce the possibility of overflow, a six-inch freeboard should be kept in the tank. Usually, after dipping, parts pass over a drip containment area and on to a drying oven. Flammable residue should not be allowed to build up on these drip areas.

Quenching Operations

Quenching operations are similar to dipping processes. The major difference is that quenching is typically done with combustible rather than flammable liquids. The principle hazard protection methods for quenching are the same as for dipping operations.

Quenching tank fires occur most frequently when the part being quenched gets hung up before being completely submerged. This allows the hot metal part to heat the combustible liquid at the surface and may result in igniting the vapors.

Other Operations

There are many other operations that typically use significant quantities of flammable liquids. For example, printing of all types uses solvents and inks fed to presses. Paint making uses numerous solvents in batch processes. Most industrial operations use flammable liquids to some extent and many use considerable quantities.

Factors Influencing Fires in Flammable Liquids

Several factors can contribute to the potential occurrence of a fire involving flammable or combustible liquids. A few of the most critical factors are personnel training, improper equipment, misuse of materials, and lack of adequate fire control measures.

An effective training program is the foundation of any flammable-liquids safety effort. Training must be provided to enable personnel to use flammable and combustible liquids in an effective and safe manner. Personnel should have a thorough understanding of the hazards associated with the materials with which they are working. They should be familiar with a liquid's degree of flammability, ease of ignition, burning characteristics, potential for explosion, and any additional hazards related to the liquid's specific method of application.

Employees must understand the importance of containing these liquids and limiting the quantities that are present in operating areas of the facility. Containment is one of the primary prevention strategies for flammable-liquid fires. If the liquid is always contained properly, a fire will not have the opportunity to start. Limiting quantities of liquids in the work area also reduces the potential for ignition and ensures that any fire will be smaller and more easily controlled.

The hazards of transfer operations and the need to constantly monitor them should be stressed because transfer operations present one of the greatest risks of fire. The transfer of the liquid from one container to another creates many opportunities for the liquid to become uncontained.

Individuals who work with or around these liquids must know the normal and emergency operating procedures. Normal operating procedures should cover the ordinary use, storage, dispensing, and any other intended usages of the flammable liquids. Emergency procedures should specify the actions to be taken and the responsibilities of personnel during spills, leaks, and fires.

One critical emergency procedure is what an individual should do if he/she is splashed with a flammable liquid. There have been far too many incidents over the years where this situation has lead to a fire in which the individual in question is killed or severely burned.

Flammable Gases

In many facilities, flammable gases are used as fuels and for industrial applications. Two primary categories of flammable gases are typically present in the industrial environment: motor fuel and industrial gases. The latter category is most often used for cutting operations.

Motor Fuel

Motor fuel cylinders (Figure 5.39) are used to provide fuel for propane-powered equipment such as forklift trucks and other vehicles used in a facility. Motor fuel cylinders may be exchanged by an outside contract service or filled on site. The contract service offers the advantage of reducing the hazards on site by eliminating transfer operations. Thus, storage and use are the only concerns. When a cylinder-filling operation is located on site, however, additional hazards have to be controlled.

Motor fuel cylinders should be stored outside in protective racks or enclosures. Filling areas for motor fuel cylinders should also be outside the building, and sources that could cause the ignition of propane vapors should be at least fifty feet from a filling area. All personnel must receive training on the proper storage, use, filling, and emergency procedures involving propane.

Welding and Cutting Gases

Gases used for welding and cutting should be stored in a secure area, and fuel gases and oxidizing gases should be separated from each other. All cylinders must be stored in a manner that minimizes the potential for physical damage to the cylinder. For example, protective valve covers should always be in place on the cylinders during storage. All gas equipment should be thoroughly inspected prior to each use. The flexible hoses and hose connections are of particular concern because they are the most likely sources of leaks.

Anytime the gas is not in use, the cylinder valve as well as the hose valves should be closed. Cylinders should be secured to the wall, a portable cart, or other transportation vehicle.

Training that covers normal use and emergency procedures is essential for all personnel who use welding and cutting gases.

FIGURE 5.39 Motor fuel cylinders.

Welding and Cutting

Welding and cutting operations present inherent risks of fire. The risks can be reduced most effectively by ensuring that welding and cutting are done in an area designated for this type of work such as a welding shop. A welding shop provides the controlled environment that is necessary for maximum safety. When welding or cutting must be done outside the welding shop, several precautions need to be taken. The most effective method of ensuring that these measures are applied is a hot work permit system. Figures 5.40 and 5.41 illustrate a simple hot work permit.

A hot work permit system is designed to provide a review of safety procedures and an inspection of the area where the welding or cutting will occur. Hot work includes any activity that introduces special ignition sources to an area. These activities may include grinding, soldering, and other items, not just welding and cutting.

For example, if welding must be done on a machine in the main plant area, the maintenance crew that will be doing the welding will request a hot work permit. If safety personnel are available in the organization, the request will usually be processed by them. However, if the organization does not have full-time safety personnel, these responsibilities will usually be handled by the maintenance manager. When the request is received, the individual will first confirm that the welding cannot be moved to the welding shop. If it cannot, the person issuing the permit will inspect the area where welding is to occur. If the area complies with the hot work

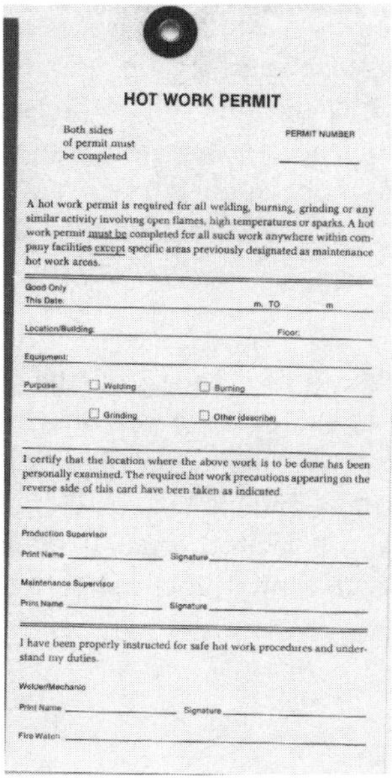

FIGURE 5.40 Hot work permit, front.

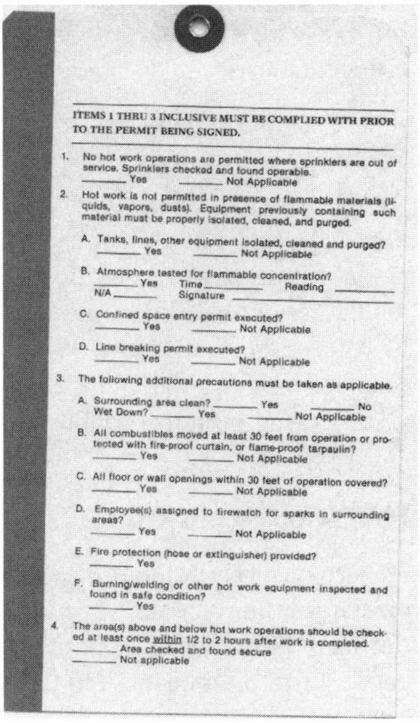

FIGURE 5.41 Hot work permit, back.

policy, a permit will be issued. A hot work permit is only valid for one shift. If the welding will take longer than one shift, a new permit must be obtained at the start of the next shift.

All welding and cutting within the facility should be covered by the hot work permit system, whether this work is done by in-house personnel or outside contractors. Agreements with contractors should specify that they will comply with all safety policies.

Combustibles within 35 feet of the welding location should be removed or protected in place with fire resistant covers. For overhead welding, covers should be suspended below the work area. Any opening passing through walls or floors should be covered.

A fire watch should be provided during the work and for at least thirty minutes after the work has been completed. The fire watch is responsible for monitoring the entire operation to ensure that a fire is not started and to quickly extinguish any small fires that may ignite. The person assigned to fire watch duties should not have other functions to perform and should maintain a constant vigilance over the operation. An extinguisher should be immediately available to the fire watch. The fire watch is also present for the safety of the welder. It is not unheard of for a welder to ignite his/her own clothing.

Machinery and Processes

All machinery and processes within the facility should be evaluated to determine the hazards and risks they present. Methods and procedures must be established to properly address these dangers. Methods of properly handling machinery and process hazards must be aimed specifically at the hazard being controlled. The general guidelines provided for hazard control earlier in this chapter can be applied to most specific situations.

Arson

Arson, which is primarily a security issue, is a major cause of fires and cannot be ignored. During hazard control activities, it is important to evaluate the potential for arson. Limiting access to the facility is one of the simplest ways to reduce the risk of arson. The probability of arson is directly influenced by many human factors. During layoffs, strikes, or other labor disputes, arson by an employee is more likely to occur than at times of good labor relations. Therefore, extra security precautions should be taken during periods of labor unrest.

HAZARD ANALYSIS

Hazard analysis is the process of evaluating the potential loss contribution of a specific hazard or an entire operation. The critical items that are identified and assessed during a hazard analysis are the likelihood that the item under consideration will cause a fire or contribute to the spread of a fire, the impact that the loss would have on operations, and the actions that can be taken to eliminate or reduce the impact of the hazard.

Hazards vary in their potential impact and probability of occurrence. Impact not only refers to the size of a loss but also to the extent of its negative influence on the operation. For example, if a fire in a particular piece of equipment would cause an entire portion of the production process to be shut down, the direct loss due to the fire may be small, but its impact would be high.

The probability assessment is the likelihood that the hazard will create a loss. Both of these factors are present to some extent in all hazards. An evaluation of the impact and probability relative to hazards allows more effective use of resources to handle hazards. The concept of impact vs. probability can be seen in the graph in Figure 5.42. Impact and probability are depicted with relative scales from zero to ten, where zero indicates little or no effect and ten indicates the most severe effect. This is not intended to be a quantifiable analysis involving detailed calculations. The position of a particular hazard on the graph is a judgment. The graph is designed to help in decision making during the process of hazard evaluation and to assist in the allocation of resources. The greatest amount of time and resources should be spent on those hazards with the highest potential for occurring and causing a negative impact. Although all hazards should be addressed, the assessment helps to establish priorities. The graph has been divided into four sections based on risk factors. Events that have a high likelihood of occurring and would have a major impact are critical risks and have the highest priority (quadrant A). Hazards which would have a major impact but are less likely to occur are placed in the second level of priority (quadrant B). Events that occur relatively frequently but whose impact on the operation is minor have third priority (quadrant C). The lowest priority is assigned to those events whose occurrence is unlikely and whose impact would be relatively minor (quadrant D). Too often in hazard control, the easy and obvious hazards get the majority of attention and resources while the difficult and less obvious hazards are ignored or assigned a low priority. This graph is designed to assist in effectively applying the limited resources available to the hazards which have the greatest potential to cause high-impact losses.

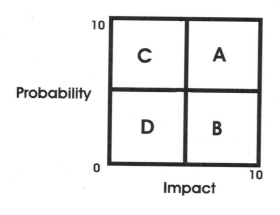

FIGURE 5.42 Probability vs. impact risk assessment.

FIRE PREVENTION PLANS

OSHA regulations require the preparation of fire prevention plans to assist with fire prevention and control activities. A properly prepared plan will identify the major fire hazards, potential ignition sources, control measures for both hazards and ignition sources, and fire protection equipment and systems. The plan must also identify the personnel in charge of fire protection equipment and systems and fuel sources. Housekeeping procedures, as well as maintenance procedures for heat-producing equipment, must be specified in the written plan. All employees must receive training which covers the fire prevention plan and their role in preventing fires. The training must also discuss fire hazards present in the workplace.

HAZARD CONTROL POLICIES AND PROCEDURES

Loss control is based on the assumption that, to an extent, all hazards can be minimized or controlled. Hazard control policies play an important part in preventing or reducing the impact of losses. The most important aspect of loss control policies is the results achieved by their guidance. A beautifully written policy is useless unless it is understood and followed. Policies should be written in a simple and straightforward manner and should be aimed at providing useful information. Policies are an aid to judgment and common sense, not a replacement for them.

A policy's primary aim should be to remind employees of proper techniques and procedures. Too frequently, policies appear to be written in a manner that indicates the writer attributed a total lack of judgment to the employees.

The fewer policies that are in place, the more likely employees will read and follow the existing policies. This also applies to the length and complexity of the policies. Several brief and apt policies are far more effective than numerous long and complex ones. Policies are tools that will become diluted with overuse. The thicker the policy manual becomes, the less likely anyone is to read it. This does not mean that issues requiring a policy should be ignored. Some operations will require more policies and additional detail in the policies. The type of operation is one major factor. For example, a chemical processing plant will require more careful and thorough policies than a cement plant because the existing hazards and the complexity of the operations at a chemical processing plant require more guidance. One effective method for solving this issue is to develop a general, company-wide policy manual and a procedure manual aimed at individual areas. If the maintenance department, for example, is the only department authorized to do cutting and welding operations in the facility, it is not necessary to cover the details of these operations in the general policy manual. The general policy might simply state that no cutting or welding is permitted in the facility without the approval of the maintenance department. The maintenance department procedure manual will discuss the requirements for safe cutting and welding in more detail.

Both policies and procedures are reminders, they are not substitutes for training. Paperwork is not a substitute for training, experience, or supervision. Using the welding example, a hot work policy and permit procedure is not enough to ensure safety. If the individual is going to be welding a container of some type, the policy or procedure would state that the container must be purged, and guidance may be provided on how this purging is to be accomplished. Without experience and training, however, the individual doing the welding will not be able to perform the job safely just by reading a policy. The welder must be trained on why the purging of the container is of critical importance, or that portion of the procedure may be skipped or be performed inadequately. In general, people will follow policies and procedures more willingly if they understand why the specific procedures are important. Training must be an integral part of the process because knowing that something is against policy is a limited motivation, but knowing that failure to follow a policy or procedure could result in the death or injury of an individual is likely to be a more effective motivation.

6 Installed Fire Protection

CHAPTER OBJECTIVES

You will be able to identify and explain:

- Construction features affecting fire protection
- Types of installed fire protection
- Basic features of each type of fire protection system
- Operation of installed fire protection

You will also be able to:

- Justify the need for installed fire protection
- Evaluate needs for installed fire protection
- Inspect installed systems

GENERAL PRINCIPLES

This chapter covers the principals of installed fire protection. Traditionally, installed fire protection systems have been considered systems designed for the detection, control, and extinguishment of fires. This limited definition eliminates many other engineering aspects of loss prevention and control. We will consider installed fire protection in two major groupings: passive systems and active systems.

This chapter will provide background and a basic understanding of a wide variety of installed fire protection systems. This short section will not make anyone an expert on any individual type of system or prepare anyone to design and engineer systems. These tasks are best left to experts in the particular systems being considered. As an individual responsible for loss prevention and control in your facility, you will need to maintain a more broad-based overall view of the use of installed fire protection in your facility.

The information in this chapter will allow you to analyze system needs, evaluate system applications, understand the operation of installed fire protection systems, and inspect these systems. The information will also allow you to converse intelligently with engineering, installation, and contract maintenance personnel.

This chapter will not be a recitation of code and regulatory requirements, but it is important to understand that those requirements will have a major impact on decisions regarding installed fire protection.

Passive Systems

Passive systems are those devices, features, and characteristics that are installed as part of a process or structure designed to avoid fire ignition, limit fire development and growth, prevent the spread of fire, and otherwise contribute to loss prevention and control efforts without any actively functioning components. An example of a passive system is a fire wall. A fire wall does not change character or operation at the time of an emergency but functions purely by its intrinsic design.

Active Systems

Active systems are components of installed fire protection that actively participate by functioning in a mechanical way at the time of an emergency. For example, a sprinkler system operates to discharge water for the purpose of control and extinguishment of a fire at the time the fire occurs. Often, when considering installed fire protection, the automatic thought is to look only at active systems, when, in fact, many useful fire protection components are passive. By overlooking passive features, we eliminate a whole category of valuable tools.

WHY INSTALLED FIRE PROTECTION IS IMPORTANT

Installed fire protection is an important aspect of effective loss prevention and control because most systems function automatically, without human intervention. Anytime human actions are involved in an operation, those phases of the operation are subject to the nuances of human behavior. Installed fire protection systems function without this human interaction. This enables the systems to perform their functions with a higher degree of reliability than any manually operated portion of the system.

Removing the need for human attention also enables these systems to provide protection to a facility twenty-four hours a day, whether the building is occupied or not. Since many large-loss fires occur from midnight to 4 a.m., this is a major advantage.

Some systems, such as dry standpipe systems, require human action. A dry standpipe system provides piping to upper floors or remote building areas that reduce the time needed to apply water to a fire emergency by removing the need to place hose lines manually. Despite the fact that they require human intervention, systems such as this one still serve a valuable function by reducing the time or resources required to perform activities.

JUSTIFYING INSTALLED FIRE PROTECTION

Whether or not an installed fire protection system is necessary is a decision that may be based on a number of criteria. The most frequent items which have an impact on the decision are: (1) cost vs. benefit, (2) life safety contribution, and (3) property protection.

The cost of installed fire protection will usually be returned within five to ten years through lower insurance premiums. This provides a net cost savings after system payback. For example, if your current fire insurance premium is $25,000 per year and the installation of a sprinkler system will drop these premiums to $5000 per year (a conservative estimate), a savings of $20,000 per year can be realized. If the installation of the sprinkler system costs $100,000 and annual maintenance costs are $2000, the system will have paid for itself in five and one-half years. If the life of the sprinkler system is 25 years (a conservative estimate), the net savings over the life of the system will be $475,000.

Another cost-savings issue is downtime. If, for example, an evaluation is made of the installation of a printing press carbon dioxide system, the cost of downtime can be used to justify the expense of installing the system. If the printing press generates a profit of $10,000 per week and the carbon dioxide system costs $25,000 to install, the cost can be justified if the system can control a fire that might put the press out of service for two and a half weeks or more. Any serious fire could easily put the press out of service for this period of time.

The protection of some items can be justified by the critical nature of the components to be protected. If a single element of an operation could create a significant hazard or cause the shutdown of many other parts of a system, protection may be warranted even if a direct cost-savings justification cannot be made. An extinguishing system on a aircraft engine, for example, is designed to save the rest of the aircraft during flight. The fire, if allowed to spread, could cause the aircraft to crash. The extinguishing agent itself will cause the engine to shut down even if it was still working at the time of discharge, but this is not the crucial point. The loss of the aircraft as a whole becomes the central issue.

Life safety contribution is another major advantage of installed fire protection. This is an important factor in motivating installed fire protection use, but it is more difficult to quantify. Many studies of fires in properties protected by installed systems have demonstrated the reduced risk to building occupants. This well established concept is reflected in most codes. The Life Safety Code® allows trade-offs such as longer travel distances to an exit in buildings with a sprinkler system. This trade-off is also another example of a cost savings. Fewer doors cost less to install and maintain.

Property protection by itself is often sufficient reason to provide installed fire protection. Even if property is insured for full replacement value, this insurance only covers part of the loss. Some items may be impossible to replace, such as prototypes, artwork, or valuable documents. Installed systems reduce the potential for losses.

FIRE WALLS

Fire walls are designed to provide separation between building areas so that a fire in one portion of a facility will not spread to others. The assumption is that even a total loss on one side of the fire wall will not spread to the opposite side.

Two of the most crucial aspects involving fire walls are (1) the initial design along with the assurance that the fire wall meets these initial specifications, and (2) the protection of openings in that fire wall. The simplest way to protect fire wall openings is to minimize the number of openings that exist in the first place. Any opening in a fire wall creates a potential problem of protection of that opening through the blocking of a fire door or some other circumstance.

Many examples of large-loss fires involving the failure of fire wall opening protection are available. One of the largest in history is the K-Mart warehouse fire. During this fire, a building divided into four segments by fire walls was completely consumed by fire due to the failure of opening protection in the fire walls.

Fire walls are rated based on the number of hours they are designed to withstand fire exposure without deteriorating or allowing the passage of fire. Typically, in an industrial setting, the most common fire walls will be four-hour walls. Fire walls extend beyond the roof line of the structure to provide protection in case of burn-through of the roof.

FIRE WALL OPENING PROTECTION

Two main types of protection for fire wall openings exist. Physical barriers such as fire doors and fire dampers are the most effective protection. Fire doors are typically horizontal doors (Figure 6.1) that are mounted on rollers and suspended from a track. These doors move horizontally to close the opening in the fire wall. Overhead rolling doors (Figure 6.2) drop vertically into place. Special fire doors may be installed in areas such as where a conveyor passes through a fire wall (Figure 6.3). Another method of protecting these openings is through the use of installed systems, for example, a water-spray system.

For openings in fire walls that are made for piping, electrical conduit, or ducts, other options are available to ensure the integrity of the fire wall is maintained, for example, fire-barrier putty (Figure 6.4). An unprotected opening through a fire wall provides a path of travel for fire to breach the wall (Figure 6.5).

FIGURE 6.1 Fire door, horizontal.

FIGURE 6.2 Fire door, overhead rolling.

FIGURE 6.3 Fire door, conveyor pass-through.

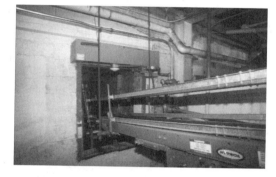

FIGURE 6.4 Fire barrier around conduit.

FIGURE 6.5 Unprotected opening through fire wall.

Fire doors are equipped with fusible links on a release mechanism that allows the door to close automatically when the heat in the area reaches a certain level (Figure 6.6). Most fire doors close by gravity through the door's own weight or through weights on a closing mechanism. It is essential that the fusible links be installed properly and that they are exposed to heat transfer from either side of the door. Doors may also be equipped with activating mechanisms attached to installed extinguishing systems. Figure 6.7 illustrates a door release attached to a carbon dioxide extinguishing system. In any area where flammable or combustible liquids may be present, curbs should be installed to prevent the passage of liquid beneath the door. In high-hazard areas, fire doors may be required on both sides of the opening.

One of the most common problems that causes fire doors not to function properly is the placement of an obstruction in the path of the door (Figure 6.8). Storage and equipment in the immediate vicinity of a fire door should be kept to a minimum. Regular checks should be made to ensure that no materials are placed in the path of the door.

Guides and bearings should be checked regularly, as should the counter-balance weights. Fire doors should not be painted. Any time that the facility is not in normal operation, fire doors should be closed manually. This is particularly important if no guard or security service is available in the plant.

FIGURE 6.6 Fire door fusible link mechanism.

FIGURE 6.7 Fire door carbon dioxide system release.

FIGURE 6.8 Blocked fire door.

ALARM SYSTEMS

Alarm systems are available in many varieties. The most basic type of alarm system would be a local manual-type system. This system is used primarily for notifying occupants of the need to evacuate the structure during an emergency. It requires manual activation by someone in the facility, usually through the use of a pull station (Figure 6.9). Figure 6.10 is an example of a pull station with a tamper-resistant cover which might be used in areas prone to frequent false alarms. Figure 6.11 is an example of a pull station for a hazardous location. After being manually activated, the alarm would then set off alarm bells, horns, strobes, or other acceptable audible or visual alarm devices (Figure 6.12). This would then alert occupants to the need for

FIGURE 6.9 Pull station.

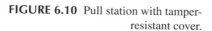

FIGURE 6.10 Pull station with tamper-resistant cover.

FIGURE 6.11 Pull station for hazardous location.

FIGURE 6.12 Alarm device.

evacuation of the structure. This alarm is not tied into any off-site facility, and the alarm does not provide any notification to the fire department.

The next level of alarm protection would be automatic detection. Automatic detection provides devices on the alarm system such as smoke and heat detectors that will detect the presence of fire and set off the alarm. A typical smoke detector is shown in Figure 6.13. A heat detector is shown in Figure 6.14. A combined smoke and heat detector is shown in Figure 6.15. This system can also be set up as a strictly

FIGURE 6.13 Smoke detector.

FIGURE 6.14 Heat detector.

FIGURE 6.15 Combination smoke and heat detector.

local alarm, where audible or visual warning devices are set off in the facility without requiring manual activation by an occupant.

Detection devices may also be associated with extinguishing systems. Figure 6.16 is a thermocouple type heat detector. A heat-actuated device (HAD) is a pneumatic system heat detector. Figure 6.17 illustrates a HAD detector. The air pressure in the contained system rises as the temperature increases. This type of system is an example of a rate-of-rise heat detection concept. There is no fixed activation temperature for this system. The system detects rapid increases in pressure associated with the rapid development of heat from a fire and activates. If slow increases in pressure occur such as gradual heating during the day, these will not activate the system. When used for installed system activation, a combination of detectors may be installed in tandem to reduce false system discharge. Flame detectors (Figures 6.18 and 6.19) may also be used. Figure 6.20 illustrates a smoke and heat detector that would require both components to detect a fire situation in order for the system to discharge. Activation of both detectors is designed to prevent false discharges of the system.

The next level of alarm would be a central-station system. This combines the elements in the previous alarm systems and provides notification to some central point at the site. Typically, this arrangement would be used where a full-time guard is on duty, and the alarm indication would be transmitted to that guard's location. This system provides local alarming for evacuation and detection and does ensure that someone in the facility is made specifically aware of the problem. At that point, the security guard could then notify the fire department, activate other in-house emergency control resources, etc. This type of system will often be equipped with an enunciator

FIGURE 6.16 Heat detector, thermo-couple type.

FIGURE 6.17 Heat detector,
HAD type.

FIGURE 6.18 Flame detector.

FIGURE 6.19 Flame detector.

FIGURE 6.20 Smoke and heat
detector.

panel (Figure 6.21). The enunciator panel shows specifically where in the facility the alarm originated.

The next level would be a remote-station alarm. Here, the alarm system is connected to an off-site monitoring station. In many areas, this would typically be the fire department itself or a contract alarm service. When the alarm detects the presence of fire, or when a manual pull-station is activated, notification is immediately transmitted to an off-site facility where the alarm is received and appropriate action is taken. If the signal goes to an outside alarm contractor, they would be given prior instructions as to whether the facility should be called to double check if the alarm

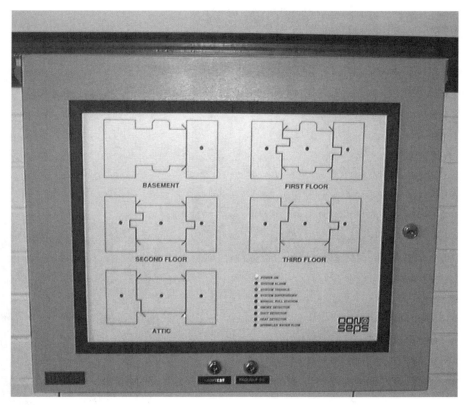

FIGURE 6.21 Alarm enunciator panel.

was false or whether the fire department should be called immediately. If the alarm signal goes to the fire department initially, the fire department would then make the appropriate dispatch of fire apparatus and fire service personnel.

STANDPIPE SYSTEMS

Standpipes are a series of piping throughout a structure that provides hose stations for occupant or fire department use at strategic locations within the facility. Figures 6.22 and 6.23 are examples of two common types of hose stations found in industrial facilities.

Standpipe systems are classified in one of three categories. A class one system (Figure 6.24) provides only a 2.5-inch (6.35 cm) hose outlet. These systems are designed for use by the fire department only. Class two systems (Figure 6.25) provide only 1.5-inch (3.8 cm) outlets and accompanying hose, usually a 100-foot (30.5 meters) section; these are primarily designed for occupant use. Class three systems (Figure 6.26) are a combination of class one and class two systems. There is a 2.5-inch (6.35 cm) outlet provided with a reducer to 1.5 inches (3.8 cm). A 1.5-inch

FIGURE 6.22 Hose station, rack.

FIGURE 6.23 Hose station, reel.

FIGURE 6.24 Standpipe hose station, class I.

(3.8 cm) hose is provided and these systems are designed for use by both occupants and fire service personnel. Fire department personnel will usually install their own hose on the outlet and rarely use the standpipe hose. Fire service hose is high quality and firefighters are familiar with the care their own hoses receive.

Standpipe systems are most valuable in facilities with very large floor area or in multi-storied facilities. The primary advantage that standpipes offer is a reduction in the time it takes to place hose lines in service. In a large facility where a fire may occur in an area 300 or 400 feet (91.4 or 121.9 meters) from an entrance point, fire service personnel would have to lay hose lines through the structure up to that point prior to being able to attack the fire. Difficult to access areas such as roofs may also be covered by standpipe systems (Figure 6.27).

FIGURE 6.25 Standpipe hose station, class II.

FIGURE 6.26 Standpipe hose station, class III.

FIGURE 6.27 Roof standpipe station.

In a facility that was equipped with standpipe hose, the firefighters could take the basic equipment they needed directly to the area of the fire, hook into the standpipe hose system, and place hose lines in service much more rapidly. This same basic principal works in structures with multiple stories.

Close in concept to standpipe systems are hose houses. A hose house is a free-standing (Figure 6.28) or wall-mounted (Figure 6.29) cabinet that contains hose, fire fighting equipment, and a water supply connection of some type, usually a fire hydrant. These devices are installed on the exterior of the building for use by fire department personnel.

AUTOMATIC SPRINKLER SYSTEMS

Automatic sprinkler systems come in several types. The main varieties are wet-pipe, dry-pipe, pre-action, and deluge. Automatic sprinkler systems are covered in OSHA regulation 1910.159 and in several NFPA codes. The primary NFPA code is NFPA 13, "Installation of Sprinkler Systems."

Appropriate sprinkler protection design includes consideration of the type of occupancy and the classification of commodities stored in the area to be protected. Occupancy types are divided into the following categories:

- Light hazard
- Ordinary hazard (group 1)
- Ordinary hazard (group 2)
- Extra hazard (group 1)
- Extra hazard (group 2)

For example, an average office area would be light hazard. An area where flammable liquids are used in moderate or substantial quantities would be an extra hazard group 2.

FIGURE 6.28 Hose house.

FIGURE 6.29 Hose house, wall mounted.

Commodity classifications are based on the fuel contribution of materials. For example, class one commodities are noncombustible materials stored directly on pallets. Class four commodities might be free-flowing polypropylene pellets. Occupancy type and commodity classification are used to help determine the design requirements of the sprinkler system. Heavier fuel loads require greater amounts of water flow if the system is to be capable of controlling a fire.

Systems used to be designed based on a system called "pipe schedule." This approach was based on a fixed number of sprinkler heads based on the size of the pipe. One-inch pipe was used for the most remote heads, and designers needed to refer to charts that indicated how many sprinklers the pipe size would support based on the hazard being protected.

Current design of sprinkler systems is based upon hydraulic calculations. Calculations are used to determine the appropriate pipe sizes necessary to achieve a specific water discharge density expressed in gallons per minute per square foot (gpm/ft^2) (liters per minute per square meter — 1 m^2). The design basis of the system should be recorded on a design plate (Figure 6.30) located at the control valve. This design plate should, at a minimum, contain the following information:

- Location of the design area or areas
- Discharge densities
- Required flow and residual pressure demand at the base of the riser
- Occupancy classification or commodity classification
- Maximum permitted storage height and configuration
- Hose stream demand included in calculations

In wet-pipe automatic sprinklers, water is provided throughout the piping of the sprinkler system at all times. The heads on a wet-pipe system are fusible devices that

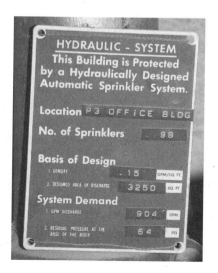

FIGURE 6.30 Hydraulic design plate.

are operated by temperature. Once the engineered temperature level is reached, the fusible element of the sprinkler head melts and allows water to flow through the head. Each sprinkler head on the system functions individually.

Dry-pipe systems are sprinkler systems with the piping normally dry. This type of system would typically be used in an unheated area. The heads and piping are basically the same as in a wet-pipe system, however air pressure is used to fill the portion of the system piping above the control valve of the sprinkler system. When a fire occurs, the heat opens the sprinkler head, allowing air to exhaust from the system and water to fill the piping.

Pre-action systems combine certain aspects of wet-pipe and dry-pipe systems. Wet-pipe systems offer the advantage of more rapid application of water the moment that the head opens. Dry-pipe systems offer the advantage of preventing the freezing of piping in unheated areas. The pre-action system combines the dry piping normally filled with air with some other fire detection method that is tied to the control valve for the water in the sprinkler system.

During a fire emergency, the detection system will detect the fire more rapidly than the fusible portion of the sprinkler heads. When the detector is set off by a fire condition, it transmits a signal back to the alarm valve which will open the valve and allow water to fill the system so that by the time the fire has reached sufficient magnitude to fuse the element of the sprinkler head, the water is already there. This reduces the delay involved in the application of water on the dry-pipe system.

Deluge systems are primarily used in high-hazard areas. The piping is very similar to regular sprinkler systems, except that the heads are open. There is no fusible element. A detection system similar to that used in a pre-action sprinkler system is used in a deluge system. When this system detects a fire condition, the sprinkler control value is opened and water will flow from all heads on the system. This is used in high-hazard areas where fires are liable to spread very rapidly, such as aircraft hangers and paint-spray booths. The problem with closed heads in these areas is that the fire would be moving too quickly and, by the time they had opened, the fire would have already passed that area.

WET-PIPE SPRINKLER SYSTEMS

Figure 6.31 shows an overall view of a wet-pipe sprinkler system and the components of the system. Figure 6.32 is a cross-section diagram of a wet-pipe sprinkler system valve in the set position. In this position, the alarm check valve is closed, separating the water in the system from the incoming water supply. The water in the system is usually maintained at an increased pressure to inhibit false trips. The alarm piping is normally open to allow water to flow to the retard chamber when the alarm check valve opens. The retard chamber is designed to retard false alarms. In a momentary water surge, the retard chamber will hold the extra flow of water without activating the alarm. A small drain in the bottom of the retard chamber allows this water to drain from the chamber in order to prepare it for another water surge. If the system activates, the flow of water will be sufficient to fill the retard chamber and allow water to proceed to the alarm circuit and water motor gong. The

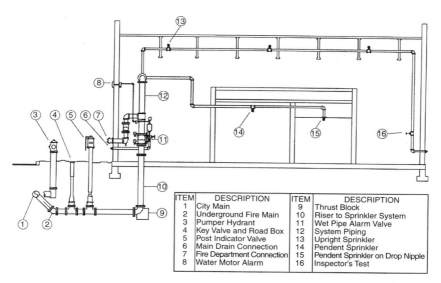

ITEM	DESCRIPTION	ITEM	DESCRIPTION
1	City Main	9	Thrust Block
2	Underground Fire Main	10	Riser to Sprinkler System
3	Pumper Hydrant	11	Wet Pipe Alarm Valve
4	Key Valve and Road Box	12	System Piping
5	Post Indicator Valve	13	Upright Sprinkler
6	Main Drain Connection	14	Pendent Sprinkler
7	Fire Department Connection	15	Pendent Sprinkler on Drop Nipple
8	Water Motor Alarm	16	Inspector's Test

FIGURE 6.31 Wet-pipe sprinkler system. (Photo courtesy of Viking Corporation.)

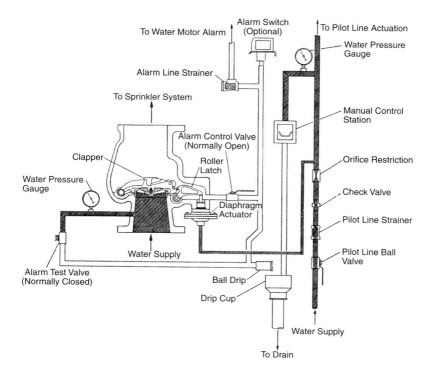

FIGURE 6.32 Cross-section of wet-pipe sprinkler valve in set position.

electrical alarm circuit is activated by a pressure switch and allows remote signaling of system activation.

Figure 6.33 is a cross-section diagram of a wet-pipe sprinkler system in the operating position. Figure 6.34 is a wet-pipe system without a control valve. Figure 6.35 is an example of a wet-pipe sprinkler system control valve. Sprinkler systems will typically be equipped with some alarm devices. The water motor gong (Figures 6.36 and 6.37) is a local alarm consisting of a bell which is operated by flowing water. The water motor gong is usually located on the outside wall near the system. Alarm components that are usually connected to a central station system include pressure switches (Figure 6.38) and flow switches (Figure 6.39) detect that the system is flowing water by either pressure or flow in the system. Another common alarm feature is a tamper switch (Figure 6.40) that will send an alarm signal when a component of the system, typically a water supply valve has been misused.

DRY-PIPE SPRINKLER SYSTEMS

Figure 6.41 shows an overall view of a dry-pipe sprinkler system and the components of the system. Figure 6.42 is a cross-section diagram of a dry-pipe sprinkler system

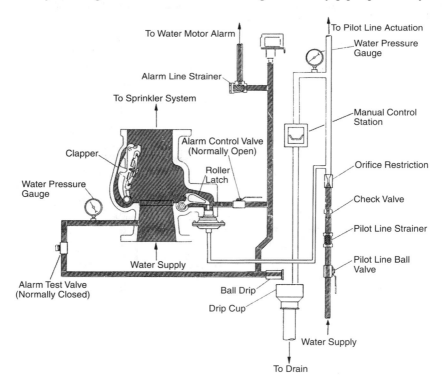

FIGURE 6.33 Cross-section of wet-pipe sprinkler valve in operating position.

FIGURE 6.34 Wet-pipe sprinkler without valve.

FIGURE 6.35 Wet-pipe sprinkler valve.

FIGURE 6.36 Water motor gong, inside building.

FIGURE 6.37 Water motor gong, out-side building

FIGURE 6.38 Pressure switch.

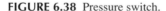
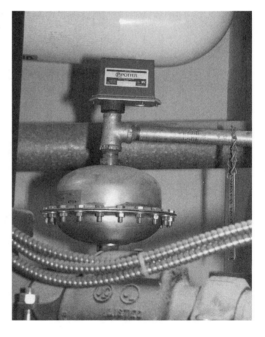

in the set position. Dry-pipe sprinkler systems are similar to wet-pipe systems. The major difference is that the piping above the alarm check valve is dry. Air pressure fills this upper piping, keeping the system free of water until a head opens and releases the pressure. A small amount of water is placed in the system through the priming cup to provide a seal at the clapper, and the air pressure in the system is

FIGURE 6.39 Flow switch.

FIGURE 6.40 Tamper switch.

maintained by an automatic compressor. The entire valve assembly is enclosed in a heated room.

When a sprinkler head on the system opens, the air exhausts and the pressure drop allows the clapper to open, permitting water flow. These systems are frequently equipped with exhausters which vent the air rapidly to the atmosphere or accelerators which pipe the air to the underside of the clapper, speeding the opening of the clapper.

Figure 6.43 is a cross-section diagram of a dry-pipe sprinkler system valve in the operating position. This diagram shows the water flow through the valve. Figure 6.44 is an example of a dry-pipe control valve. A pre-action system valve is shown in Figure 6.45. A deluge sprinkler valve is shown in Figure 6.46.

Sprinkler Heads

Sprinkler heads are available in many different styles and temperature ratings. The two most common types are pendent (Figure 6.47) and upright (Figure 6.48). The

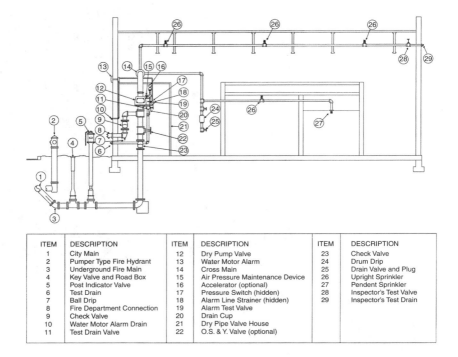

ITEM	DESCRIPTION	ITEM	DESCRIPTION	ITEM	DESCRIPTION
1	City Main	12	Dry Pump Valve	23	Check Valve
2	Pumper Type Fire Hydrant	13	Water Motor Alarm	24	Drum Drip
3	Underground Fire Main	14	Cross Main	25	Drain Valve and Plug
4	Key Valve and Road Box	15	Air Pressure Maintenance Device	26	Upright Sprinkler
5	Post Indicator Valve	16	Accelerator (optional)	27	Pendent Sprinkler
6	Test Drain	17	Pressure Switch (hidden)	28	Inspector's Test Valve
7	Ball Drip	18	Alarm Line Strainer (hidden)	29	Inspector's Test Drain
8	Fire Department Connection	19	Alarm Test Valve		
9	Check Valve	20	Drain Cup		
10	Water Motor Alarm Drain	21	Dry Pipe Valve House		
11	Test Drain Valve	22	O.S. & Y. Valve (optional)		

FIGURE 6.41 Dry-pipe sprinkler system. (Photo courtesy of Viking Corporation.)

upright head is installed on the top side of the sprinkler piping and the pendent head is installed on the bottom side of the sprinkler piping. The components of a sprinkler head are shown in Figure 6.49.

Sprinkler heads are also available in recessed styles (Figure 6.50), concealed styles, and sidewall. Many other varieties are also available. In areas such as storage racks where the sprinkler may be subject to physical damage, a cage is installed around the sprinkler (Figure 6.51). Sprinkler heads may also be coated with wax to protect them from corrosive vapors in the environment. The wax melts off in the event of a fire and the sprinkler head can then function normally. Sprinklers that must be installed with a significant amount of space above them may be equipped with a heat retainer (Figure 6.52). This heat retainer collects heat in the area of the head to avoid slowing down the response of the fusible element.

Sprinkler heads are rated based on the temperature at which they operate. The standard sprinkler head fuses at 135–170°F (57.2–76.6°C) and has a maximum acceptable ceiling temperature of 100°F (37.7°C). Higher-temperature heads are available for installation in areas where ambient temperatures exceed 100°F (37.7°C).

Most sprinklers have a standard discharge orifice of 0.5 inch (1.27 cm) which provides a flow of approximately 22 gallons per minute (83 liters/minute) at a pressure of 15 pounds per square inch (103 kilopascals). Large orifice heads are available for use when flows need to be higher.

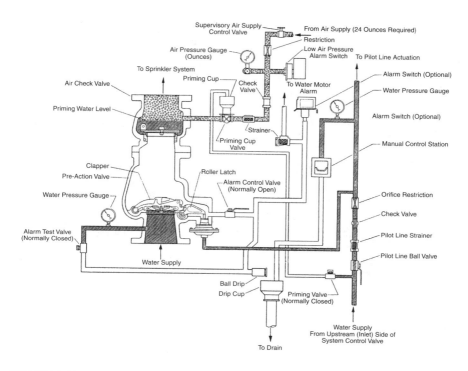

FIGURE 6.42 Cross-section of dry-pipe sprinkler valve in set position.

Extra sprinkler heads must be maintained on site so that used heads may be replaced immediately. These extras should include all of the types used in the facility. Typically, the extra heads are stored in a cabinet (Figure 6.53) mounted near the sprinkler control valve.

SPRINKLER SYSTEM INSPECTION AND MAINTENANCE

For sprinkler systems to function properly at the time of a fire, they must be inspected and maintained. All valves on the system should be inspected weekly. The most common valves are outside screw and yoke (OS & Y) (Figure 6.54) and butterfly (Figure 6.55) because whether these valves are open or closed can be determined at a glance. Pressures should also be checked weekly. In dry-pipe systems, the air and heating mechanisms should be inspected weekly.

At least once annually, the overall function of the system should be evaluated by activating the system for a test. This is done by opening an inspector's test valve (Figure 6.56). This valve is located on the system at a point remote from the alarm check valve and is designed to simulate the opening of a sprinkler head. Piping to the system can be checked by conducting a main drain flow test. The flow test can help evaluate the condition of piping flowing to the sprinkler system. This piping may be-

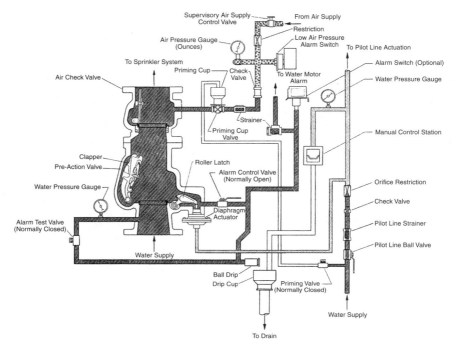

FIGURE 6.43 Cross-section of dry-pipe sprinkler valve in operating position.

FIGURE 6.44 Dry-pipe sprinkler valve.

come obstructed by foreign material, scale, rust, and microbial action. Insurance company representatives will often perform these annual tests as part of the company's insurance service and in order to satisfy themselves that the system is functioning properly. It is often easier to contract with a sprinkler system installation company for these annual evaluations than to have in-house personnel stay current on proper procedures if the insurance company does not provide this service.

FIGURE 6.45 Pre-action sprinkler valve.

FIGURE 6.46 Deluge sprinkler valve.

FIGURE 6.47 Sprinkler head, pendent.

FIGURE 6.48 Sprinkler head, upright.

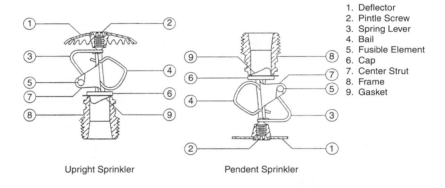

1. Deflector
2. Pintle Screw
3. Spring Lever
4. Bail
5. Fusible Element
6. Cap
7. Center Strut
8. Frame
9. Gasket

Upright Sprinkler   Pendent Sprinkler

FIGURE 6.49 Sprinkler head components.

FIGURE 6.50 Sprinkler head, recessed.

FIGURE 6.51 Sprinkler head in cage.

FIGURE 6.52 Sprinkler head with heat retainer.

FIGURE 6.53 Extra sprinkler head cabinet.

WATER-SPRAY SYSTEMS

Water-spray systems (Figure 6.57) are special application systems that apply water in a fine spray. A water-spray system is similar to a deluge system in that the sprinkler heads are all open (Figure 6.58). Water-spray systems are used in areas where

FIGURE 6.54 Outside screw and yoke (OS & Y) valve.

FIGURE 6.55 Butterfly valve.

FIGURE 6.56 Inspector's test valve.

three-dimensional protection and higher-velocity streams are required. A sprinkler system discharges the water in a downward direction, and gravity is the primary method of transporting the water to the fire. Water-spray systems allow forced application of water. These systems may be found protecting, for example, large transformers, propane storage tanks, and firewall openings.

Foam Systems

Foam systems are specialized systems used in areas where flammable-liquid fires are expected. Foam systems come in two main varieties: low-expansion foam and high-expansion foam. Low-expansion foam systems can use any of the common varieties of low-expansion foam discussed in the segment on extinguishing agents. Low-expansion foam may be discharged through a special type of sprinkler head (Figure 6.59) or through monitor nozzles (Figure 6.60). These foams may also be made available for manual application through standpipe systems. High-expansion foam systems are speciality systems that apply high-expansion foam. High-expansion foam is typically applied from a foam generator (Figures 6.61 and 6.62) mounted at roof level. Foam systems are used in high hazard areas such as aircraft hangars and flammable-liquid storage areas.

Foam storage (Figure 6.63) is an additional component of the system. Foam is made by combining a foam concentrate with water. This process is referred to as proportioning and is typically done either in-line with an eductor, through a process called balance pressure proportioning, or with an around-the-pump proportioner.

FIGURE 6.57 Water spray system.

FIGURE 6.58 Water spray system head.

FIGURE 6.59 Air aspirating foam sprinkler head.

FIGURE 6.60 Monitor nozzle. (Photo courtesy of Ansul Incorporated.)

FIGURE 6.61 High expansion foam generator. (Photo courtesy of Ansul Incorporated.)

FIGURE 6.62 High expansion foam generator installed in roof.

WATER SUPPLY SYSTEMS

An effective water supply is essential for a water-based extinguishing system. All water supplies have several basic components: a source of water, piping to transfer water, and a method for moving water. Water supply system capacities are determined by the required fire flow and the duration that the flow must be maintained. For example, if an installed system may need to provide flows of 1000 gpm (3785 lpm) for one hour, the water supply would need a capacity of 60,000 gallons (227,100 liters).

The ideal arrangement for water supply systems is to provide two sources of water. This provides a backup should one source be out of service. Typically, one source is a city water supply and the other is storage on site. In areas where a city water supply is not available, it is possible to use two separate storage arrangements.

Storage for the water supply will most commonly be one of the following: elevated tank, suction tank, reservoir, or pressure tank. An elevated-tank storage (Figure 6.64) arrangement offers the advantage of gravity-induced head pressure. A 100-foot-high (30.5 meters) tank would provide an outlet pressure of 43 psi (296 kilopascals) at the base simply through the action of gravity. Depending on specific system needs, a gravity tank can be an acceptable water supply without a fire pump. Suction tanks provide a less expensive storage arrangement than elevated tanks but need a fire pump. Figure 6.65 illustrates a typical suction tank and pump house arrangement. Pressure tanks use air pressure to expel the water, much like a pressurized water fire extinguisher only on a much larger scale. Pressure tanks are typically used in small systems. All of the three storage arrangements just discussed require inspection and maintenance on a regular basis. In areas where temperatures may fall below freezing, a method for heating the water is also required. The use of a reservoir (Figure 6.66) eliminates inspection and maintenance almost completely. Freeze protection is not necessary. The biggest disadvantage of the reservoir is the space it consumes and the cost of construction.

Piping is needed to move the water from storage to systems for application during a fire. The water main system must have adequate flow capacity to supply all the demands. The size of water mains is determined by required flow. Ideally, water

FIGURE 6.63 Foam storage tank. (Photo courtesy of Ansul Incorporated.)

FIGURE 6.64 Elevated tank.

FIGURE 6.65 Suction tank and
pump house.

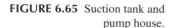

mains should be set up to form a loop (Figure 6.67) because this allows service to be provide to most areas of the system even if part of the system is damaged or shut down for maintenance. Individual systems can be fed off the loop.

Valves on fire protection water supply systems are usually post-indicator valves (PIV) (Figure 6.68) and wall indicator valves (Figure 6.69). The valves resemble posts and the window indicates an open or shut position. These valves are typically locked in the open position and/or equipped with tamper switches.

FIGURE 6.66 Reservoir.

Water supply systems are usually provided with a fire pump to boost pressure and flow in the water system. From the water supply storage device the water is piped to the pump. The main fire pump may be powered in a number of ways, the most common of which are diesel engine (Figure 6.70) and electric motor driven pumps (Figures 6.71 and 6.72). The pumps are provided with a controller that is designed to start the fire pump automatically at a predetermined drop in pressure within the system. This controller or another device should provide a remote alarm to signal that the fire pump has started. A small pump, called a jockey pump (Figure 6.73), is provided to maintain pressure in the system and prevent the main pump from being started needlessly. Small drops in pressure caused by leakage, for example, will only activate the jockey pump.

Water-based installed systems will also be provided with a fire department connection (FDC) (Figure 6.74) which allows the fire department to pump water into the system from an external source.

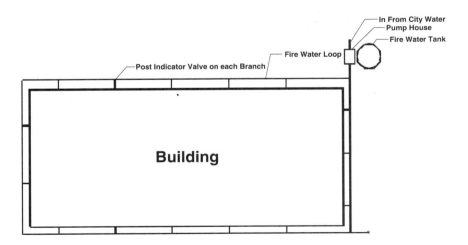

FIGURE 6.67 Fire protection loop system.

FIGURE 6.68 Post indicator valve (PIV).

FIGURE 6.69 Wall PIV.

FIGURE 6.70 Fire pump, diesel.

FIGURE 6.71 Fire pump, electric, horizontal shaft.

FIGURE 6.72 Fire pump, electric, vertical shaft.

FIGURE 6.73 Jockey pump.

The water supply system will also include hydrants. These may be part of the site's system or may be public. Common varieties include the dry barrel hydrant (Figure 6.75), wet barrel hydrant (Figure 6.76), and wall hydrant (Figure 6.77). The dry barrel hydrant has the valve below grade where the hydrant is connected to the water main. All of the outlets on the hydrant are controlled by the valve stem on top of the hydrant. These hydrants are the most common and are not subject to freezing. In areas where freezing is not a possibility, wet barrel hydrants may be used. These hydrants are equipped with control valves for each outlet. Wall hydrants are used to

FIGURE 6.74 Fire department connection.

FIGURE 6.75 Dry barrel hydrant.

FIGURE 6.76 Wet barrel hydrant.

FIGURE 6.77 Wall hydrant.

provide water for manual fire fighting in areas of industrial sites where it would be impractical to place an ordinary hydrant.

WATER SUPPLY SYSTEM INSPECTION AND MAINTENANCE

Water supply is critical to water-based systems and must be maintained in good condition at all times. Supply tanks must be provided with a means for refilling and tank levels should be checked regularly. Heating mechanisms should be checked at least weekly during cold weather and serviced annually prior to the heating season. Piping should be checked by flow-testing at least annually. Fire pumps need to be run weekly to ensure that the starting mechanism functions properly. Pumps and drive engines or motors need to be thoroughly serviced every year. Pumps should be flow-tested annually to ensure that acceptable performance is being maintained. A test head (Figure 6.78) is provided for this purpose.

CARBON DIOXIDE SYSTEMS

Carbon dioxide systems are available in two primary types: local-application systems and total-flooding systems. They are also available in two primary types of agent supply set ups: high-pressure and low-pressure. Carbon dioxide systems are typically used on equipment and devices that would have flammable or combustible liquids in use but which would be severely damaged by the use of dry chemicals. Examples would be printing presses or metal rolling equipment. Carbon dioxide is considerably less expensive than halon. Carbon dioxide systems include the following components: storage, activation, and distribution. The use of carbon dioxide systems should be avoided in any areas that cannot be quickly evacuated, areas where pyrophoric materials are used, and areas where reactive metals are used.

Low-pressure systems (Figure 6.79) have pressures up to 300 psi (2069 kilopascals), are refrigerated to 0°F (−17.8°C), and are filled to a 90 to 95% filling density. High-pressure systems (Figure 6.80) operate at pressures of 850 psi (5863 kilopascals) and are maintained at a normal ambient temperature of 70°F (21°C). They are filled to a 60 to 68% filling density. Heat (Figure 6.81) or flame (Figure 6.82) sensors typically activate the system, which then discharges carbondioxide through strategically placed nozzles. To warn people in the area to evacuate, systems are equipped with pre-discharge alarms (Figure 6.83). These alarms should provide sufficient time to escape the immediate area of system coverage prior to the discharge

FIGURE 6.78 Fire pump test head.

FIGURE 6.79 Carbon dioxide system low-pressure storage. (Photo courtesy of Ansul Incorporated.)

FIGURE 6.80 Carbon dioxide system high-pressure storage.

FIGURE 6.81 Carbon dioxide system heat detector.

FIGURE 6.82 Carbon dioxide flame detector and nozzle.

FIGURE 6.83 Carbon dioxide system predischarge alarm.

FIGURE 6.84 Carbon dioxide system nozzle.

of carbon dioxide. These systems function by displacing the oxygen in the area of coverage and pose a significant risk to personnel that remain in the area.

Carbon dioxide system nozzles (Figure 6.84) are similar to small versions of the discharge horn on a portable carbon dioxide fire extinguisher. In local application systems, the nozzles will be directed toward the hazard being protected. Nozzle position is important for the system to function properly. In total flooding systems, nozzles will typically be located on the ceiling. In these systems, retaining the carbon dioxide within the area is critical and there will usually be door (Figure 6.85) and vent (Figure 6.86) closers.

The systems may also be activated manually; a manual trip is shown in Figure 6.87. Systems are also available that are manually activated only. These systems (Figure 6.88) function much like portable fire extinguishers except the discharge of carbon dioxide is directed through fixed piping to a specific hazard. Carbon dioxide systems may also be equipped with hand lines (Figure 6.89) for use in manual fire fighting.

FIGURE 6.85 Carbon dioxide system door closer.

FIGURE 6.86 Carbon dioxide system vent closer.

FIGURE 6.87 Carbon dioxide system manual trip.

Figures 6.90 and 6.91 show carbon dioxide systems discharging. Carbon dioxide will accumulate at the floor level and in low lying areas (Figure 6.92). These areas may remain oxygen deficient for an extended period of time after the system has discharged.

FIGURE 6.88 Manual carbon dioxide system.

FIGURE 6.89 Carbon dioxide system hand line.

HALON SYSTEMS

Halon systems are gaseous-agent systems typically used in computer rooms or electronic control areas. Usually they are designed as total-flooding systems. Total flooding systems are systems which disburse a gas throughout the entire area being protected. The agent is stored in cylinders or spheres (Figure 6.93) and is discharged through strategically placed nozzles (Figure 6.94).

These systems are usually activated by smoke detectors on a two-zone system. The two-zone system requires that a detector in each zone sense smoke before tripping the system. This arrangement reduces the opportunity for false discharges of the system. In computer rooms, where many of these systems are installed, smoke detectors will be installed below the raised floor (Figure 6.95).

Halon is an effective agent, with the advantage that it leaves no residue or mess and can be used to protect delicate equipment such as computers. At proper design concentrations it poses little risk to personnel in the protected area. The biggest

FIGURE 6.90   Carbon dioxide system discharge.

FIGURE 6.91 Carbon dioxide system discharge.

FIGURE 6.92   Carbon dioxide accumulated at floor level.

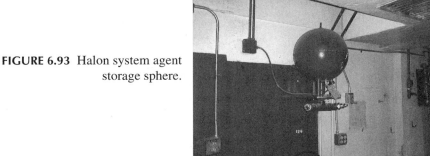

FIGURE 6.93 Halon system agent storage sphere.

FIGURE 6.94 Halon system nozzle.

disadvantage is the high cost of the agent. Due to the environmental regulations discussed in Chapter 2, these systems are not typically being installed anymore but are permitted to remain in service at existing installations.

HALON SUBSTITUTES

Most of the halon substitute agents discussed in Chapter 2 are available for use in installed systems. From a practical perspective, there is little difference in the basic components when compared to halon systems. The extinguishing agent is stored in cylinders (Figure 6.96) and is applied through strategically located nozzles (Figure 6.97).

FIGURE 6.95 Smoke detector
below raised floor.

FIGURE 6.96 INERGEN® sys-
tem. (Photo courtesy of Ansul
Incorporated.)

FIGURE 6.97 INERGEN® system nozzle.
(Photo courtesy of Ansul Incorporated.)

DRY CHEMICAL SYSTEMS

Dry chemical systems (Figure 6.98) are local-application systems which provide dry chemical agent of any of the common varieties to a specific hazard such as a dip tank. The system is activated by a heat sensor which trips the system and allows the dry chemical to discharge through strategically placed nozzles. Dry chemical systems

FIGURE 6.98 Dry chemical system.

FIGURE 6.99 Flammable vapor sensor.

provide rapid control of flammable-liquid fires, but the chemical can create a cleanup problem. Areas where delicate equipment is located should generally not be protected with dry chemical systems.

FLAMMABLE VAPOR DETECTION SYSTEMS

Flammable vapor detection systems are typically used to control processes that have the potential to generate explosive atmospheres. Two stages of control are usually provided. First, at typically 25% of the LEL, the system will provide a warning alarm. Second, at 50% LEL, the system will automatically shut down the protected process. These set points for the alarm and automatic shut-down may be selected by the user.

These systems may be electronic or combustion based. In electronic systems, sensors (Figure 6.99) are mounted in locations around or within the process and send electronic signals to control panels, alarms, and possibly process controls. In combustion based systems, the control panel (Figure 6.100) of the system uses a small flame, and sample atmosphere from the process being monitored is drawn to the control through tubing (Figure 6.101).

FIGURE 6.100 Flammable vapor sensor control.

FIGURE 6.101 Flammable vapor draw tube.

EXPLOSION VENTING AND SUPPRESSION SYSTEMS

Systems to protect from explosions are important in many industrial operations. Explosions are possible in operations using flammable liquids, flammable gases, and many dusts. Explosions in this context are actually deflagrations. In deflagrations, the pressure wave propagating out from the initial point of ignition moves at speeds less than the speed of sound. A detonation creates pressure waves that move in excess of the speed of sound. The same things required for a fire — fuel, heat, oxygen, and a chemical reaction — are also required for an explosion. Explosive forces create a much more damaging potential if confinement is added to this mix.

If the circumstances of the process permit, venting is one of the most effective ways to handle an explosion without extensive damage to the process or personnel in the immediate area. Venting is accomplished by introducing a weak component into a process system. This weak portion of the system will fail immediately during an internal explosion and vent the pressure increase and fire in a controlled direction.

Explosion suppression systems are typically utilized in areas where flammable vapors or gases, explosive dusts, or other readily explosive materials are present. These systems function so rapidly that they can detect the beginning stages of an explosion, discharge agent into the area, and effectively abort an explosion prior to a dangerous build up of pressure. Detection is typically pressure increase or flame based. Agent canisters are placed strategically on the process being protected. Dry chemical, gaseous, and vaporizing liquid agents are used in these systems.

INSTALLED SYSTEM IMPAIRMENTS

Impairments are when systems are out of service due to either a problem on the system or maintenance. Impairment periods are particularly dangerous times and should be planned for well in advance. Only one system should be taken out of service at a time. Once work is started on a system it should continue until the system is back in service. This requires a thorough evaluation of the work to be done so that all necessary parts and equipment are available before the system is taken out of service.

During impairments, extra precautions should be taken in the areas affected by the impairment. Restrictions on smoking, on the use of flammable liquids, and on welding or cutting are some examples of these precautions. The insurance company and local fire department should be notified prior to the impairment and after the system is returned to service.

INSPECTION AND MAINTENANCE OF SYSTEMS

All installed fire protection systems require regular inspection and maintenance. Unless your organization is large and is willing to obtain appropriate training for an in-house person, it is more effective to contract for inspection and maintenance services with the company that installed the system. If an in-house person will be used, he/she should receive training on all systems, preferably from the manufacturer of the system.

No matter who does the inspection and maintenance, a record should be maintained indicating all inspection and maintenance activities, when they were completed, who did the work, and what was done.

EVALUATE SYSTEM NEEDS

The first principle when evaluating the need for installed fire protection systems is that it is better to have systems than not to have systems. These systems provide a high level of reliable fire protection. They have been proven over time to reduce losses, so they are generally an excellent addition to the overall loss-control program.

There are a few situations in which the cost of a system or the difficultly in providing installed fire protection may warrant not considering a system. For example, a small outside storage building, well away from other buildings, may not be worth the cost to protect. But in most situations, protection is needed and the primary purpose of the systems evaluation becomes determining which type of system will provide the most effective protection.

The hazard being protected should be the prime criterion for systems evaluation. The system needs to provide effective fire protection. Once a selection of effective systems has been identified, the complexity and cost of the systems need to be considered. Generally, simple is better than complex, and the most basic system that will address the hazard is usually best.

Any evaluation of installed fire protection should consider long-term costs as well as initial costs. The annual inspection and maintenance of the system must be considered when evaluating which system will best meet the specific need. For example, a carbon dioxide system may cost more for initial installation than a halon system for a particular hazard, but if frequent discharges can be expected, such as in a metal rolling mill, the carbon dioxide system is cheaper to recharge and may cost less over the life of the system.

Help will probably be needed to conduct a comprehensive evaluation of the need for an installed system and to make a decision as to the type of system necessary. Do not consider the advice of the contractor who will install the system as your only source of information. The vast majority of contractors are honest and will not intentionally provide poor advice, but an unbiased second opinion is always a good idea. The insurance company, the local fire department, or a consultant can provide input that may save money and problems.

7 Portable Fire Extinguishers

CHAPTER OBJECTIVES

You will be able to identify and explain:

- Types of extinguishers
- Performance characteristics of extinguishers
- Principles of selection
- Principles of placement
- Principles of use
- Inspection and maintenance requirements

You will also be able to:

- Select extinguishers
- Place extinguishers
- Use extinguishers
- Develop employee training concerning extinguishers
- Develop an extinguisher inspection program
- Inspect extinguishers
- Develop an extinguisher maintenance program

TYPES AND PERFORMANCE CHARACTERISTICS OF EXTINGUISHERS

Many types of fire extinguishers are currently available, and each one has certain advantages and disadvantages. For that reason, it is important to understand the capabilities of each type of extinguisher when selecting one to provide protection for a specific hazard or area.

WATER EXTINGUISHERS

Water extinguishers are available in two types: the pump-type extinguisher discharges the water through a pump operated by the user, and the pressurized extinguisher uses air pressure to discharge the water.

Pump-type extinguishers are not commonly used. Although some operations may still have and use them, pressurized water extinguishers will most often be employed in situations where water is the appropriate extinguishing agent. Pressurized

FIGURE 7.1 Pressurized water extinguisher. (Photo courtesy of Amerex Corporation.)

water extinguishers (Figure 7.1) have a capacity of 2.5 gallons (9.5 liters) and are suitable for use on class A fires only. They have a range of approximately 30 feet (9.2 meters) when initially discharged and a discharge time of about one minute. They are pressurized with 100 psi (690 kilopascals) of compressed air.

Water extinguishers have several advantages. In many situations, water is easy to clean up and causes little or no additional damage. The extinguisher is inexpensive to recharge and can easily be returned to service by in-house personnel with limited training. In addition, the soaking ability of water is a major advantage in fighting class A fires.

There are also a number of disadvantages to water extinguishers. For example, these units are only effective on class A fires and can be dangerous if used on any other class of fire. The extinguishers are also relatively heavy. In any area that is not continuously heated, the unit is subject to freezing. Antifreeze can be added to prevent freezing, but that makes maintenance more difficult.

A specialized type of water-based extinguisher is the water mist (Figure 7.2) variety. These extinguishers use distilled water and a nozzle that discharges a fine mist to achieve a Class A & C rating for the extinguisher. Water mist extinguishers are available in 1.75 and 2.5 gallon (6.6 and 9.5 liter) sizes. Their range is 10 to 12 feet (3 to 3.7 meters) and discharge time is between 72 and 80 seconds.

FOAM EXTINGUISHERS

Portable foam extinguishers are modified pressurized water extinguishers and are also available in two types. The premix foam extinguisher (Figure 7.3) contains a

FIGURE 7.2 Water mist extinguisher. (Photo courtesy of Amerex Corporation.)

FIGURE 7.3 Foam extinguisher. (Photo courtesy of Amerex Corporation.)

mixture of water and foam concentrate within the extinguisher. An aerating nozzle is typically used to make foam from the foam solution as it is discharged. The cartridge-type foam extinguisher contains water only in the body of the extinguisher. At the time of discharge, the water flows through a cartridge containing pellets that make foam when mixed with the water. The other construction features of the extinguisher are similar to a water extinguisher.

Foam extinguishers have a capacity of 2.5 gallons (9.5 liters) and are suitable for use on class A and B fires. They have a range of slightly less than the 30 feet (9.2 meters) typical of water extinguishers, and their discharge time is about one minute. They are pressurized with 100 psi (689 kilopascals) of compressed air.

Both foam extinguisher types have several advantages. In many situations, foam is easy to clean up and causes little or no additional damage. The extinguisher is relatively inexpensive to recharge and can be returned to service by in-house personnel with proper training. Class B fires can be controlled with foam extinguishers. Unlike any other type of portable fire extinguisher, foam extinguishers can cover Class B spills with foam to help prevent ignition.

Foam extinguishers also have a number of disadvantages. For example, these units are only effective on class A and B fires and can be dangerous if used on class

C or D fires. The extinguishers are relatively heavy. In any area that is not continuously heated, the unit is subject to freezing. Antifreeze can be added to prevent freezing, but that makes maintenance more difficult.

Foam extinguishers are also available in wheeled units (Figure 7.4), and their most common size is 33 gallons (125 liters).

FIGURE 7.4 Foam wheeled extinguisher. (Photo courtesy of Ansul Incorporated.)

DRY CHEMICAL EXTINGUISHERS

Portable dry chemical extinguishers are available in two types. The stored pressure extinguisher (Figure 7.5) contains a dry chemical agent and pressurizing gas, usually nitrogen, within the extinguisher. The valve head and gauge are shown in Figure 7.6, which shows the tamper seal and pin. This type of extinguisher is available in many sizes ranging from 2.5 to 30 pounds (1.1 to 13.6 kilos). Ten-pound (4.5 kilo) units are the most commonly used in building fire protection. Small units of 2.5 pounds (1.1 kilos) (Figure 7.7) are commonly placed on forklifts and in vehicles. Discharge times vary from 8 to 20 seconds, and the discharge range is from 5 to 30 feet (1.5 to 9.2 meters).

The cartridge operated dry chemical extinguisher (Figure 7.8) contains dry chemical agent in the body of the extinguisher and a separate cartridge of pressurizing gas. When the extinguisher is needed, its main body is pressurized from this cartridge. The top of the extinguisher including the charging handle is shown in Figure 7.9. Figure 7.10 shows the extinguisher with the cartridge guard removed so the charging cylinder is visible. Figure 7.11 shows the discharge nozzle for a cartridge operated extinguisher. The sizes and discharge times and ranges are similar to stored pressure extinguishers.

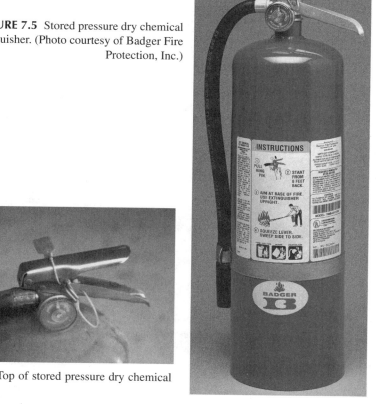

FIGURE 7.5 Stored pressure dry chemical extinguisher. (Photo courtesy of Badger Fire Protection, Inc.)

FIGURE 7.6 Top of stored pressure dry chemical extinguisher.

Both types are available with either major class of dry chemical agent: regular dry chemical or multipurpose dry chemical. Regular dry chemical may be used on class B and C fires. Multipurpose dry chemical is effective on class A, B, and C fires.

Dry chemical extinguishers (Figure 7.12–7.14) are also available in wheeled units whose sizes range from 50 to 350 pounds (22.6 to 159 kilos). Their discharge range varies from 15 to 45 feet (4.6 to 13.7 meters), and their discharge time is from 30 to 150 seconds.

Both types of dry chemical extinguishers have several advantages. They offer rapid fire control, particularly for class B fires. The units are not subject to freezing, so they can be placed outside or in unheated areas. The cartridge operated extinguisher is relatively inexpensive to recharge and can be returned to service by in-house personnel with proper training. Stored pressure extinguishers, however, cannot be recharged in-house without costly, specialized equipment and more extensive personnel training.

FIGURE 7.7 Stored pressure dry chemical extinguisher, small unit. (Photo courtesy of Badger Fire Protection, Inc.)

FIGURE 7.8 Cartridge dry chemical extinguisher. (Photo courtesy of Ansul Incorporated.)

FIGURE 7.9 Top of cartridge operated dry chemical extinguisher.

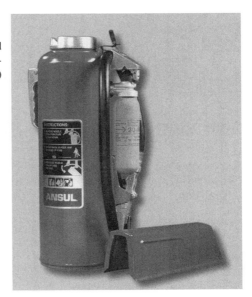

FIGURE 7.10 Cartridge guard removed showing charging cylinder. (Photo courtesy of Ansul Incorporated.)

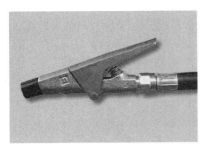

FIGURE 7.11 Discharge nozzle for cartridge operated extinguisher. (Photo courtesy of Ansul Incorporated.)

FIGURE 7.12 Wheeled dry chemical extinguisher, small. (Photo courtesy of Amerex Corporation.)

FIGURE 7.13 Wheeled dry chemical extinguisher, medium. (Photo courtesy of Amerex Corporation.)

FIGURE 7.14 Wheeled dry chemical extinguisher, large. (Photo courtesy of Ansul Incorporated.)

There are also a number of disadvantages to dry chemical extinguishers. For example, regular dry chemical is not effective on class A fires. Multipurpose dry chemical, although effective and approved for use on class A fires, is not as effective as water. Also, both types can make a mess that is difficult and costly to clean up. In situations involving delicate electronic equipment, the cleanup costs can be higher than the initial fire damage.

CARBON DIOXIDE EXTINGUISHERS

Carbon dioxide (CO_2) extinguishers contain carbon dioxide forced into its liquid state by pressure. This type of extinguisher is available in several sizes from 2.5 to 20 pounds (1.1 to 9 kilos). Discharge times range from 8 to 30 seconds, and the discharge range is from 5 to 8 feet (1.5 to 2.4 meters).

Small carbon dioxide units have a discharge horn attached by tubing (Figure 7.15), whereas larger units have a discharge horn at the end of a flexible hose (Figure 7.16). Carbon dioxide extinguishers are also available in wheeled units (Figure 7.17 and 7.18). Their sizes range from 50 to 100 pounds (22.6 to 45.4 kilos). Their discharge range varies from 3 to 10 feet (0.9 to 3 meters), and their discharge time is from 10 to 30 seconds.

FIGURE 7.15 Small carbon dioxide extinguisher. (Photo courtesy of Ansul Incorporated.)

FIGURE 7.16 Large carbon dioxide extinguisher. (Photo courtesy of Ansul Incorporated.)

FIGURE 7.17 Wheeled carbon dioxide extinguisher.

FIGURE 7.18 Wheeled carbon dioxide extinguisher. (Photo courtesy of Amerex Corporation.)

Carbon dioxide extinguishers have several advantages. They can be used on class B or C fires, and since carbon dioxide does not leave a residue, cleanup is not a problem. The units are not subject to freezing, thus they can be placed outside or in unheated areas. There are also a number of disadvantages to carbon dioxide extinguishers. The units are not suitable for class A fires. Static electricity can build up during discharge of the extinguisher and could feasibly ignite an explosive atmosphere or damage sensitive electronic circuits. These extinguishers cannot be recharged in-house without the purchase of specialized equipment and personnel training.

Halon Extinguishers

Halon extinguishers (Figure 7.19) contain a halogenated hydrocarbon agent forced into a liquid state by pressure. The most common halons for portable extinguishers are Halon 1211 (bromochlorodifluoromethane) and Halon 1301 (bromotrifluoromethane). This type of extinguisher is available in several sizes from 1.5 to 22 pounds (0.68 to 9.9 kilos). Discharge times range from 8 to 30 seconds. Discharge range is from 9 to 15 feet (2.7 to 4.6 meters). These extinguishers are also available in 150-pound (68 kilo) wheeled units (Figure 7.20) which have discharge times of 30 to 35 seconds and a range of 10 to 18 feet (3 to 5.5 meters).

FIGURE 7.19 Halon extinguisher. (Photo courtesy of Ansul Incorporated.)

FIGURE 7.20 Wheeled Halon extinguisher. (Photo courtesy of Ansul Incorporated.)

Halon extinguishers have several advantages. They can be used on class B or C fires, and larger units are also approved for use on class A fires. Halon leaves no residue, so cleanup is not a problem. The units are not subject to freezing, so they can be placed outside or in unheated areas. Generally, they offer more effective fire control than carbon dioxide extinguishers.

Halon extinguishers also have a number of disadvantages. For example, the smaller units are not suitable for class A fires, and halon is expensive. Recharge costs can approach or exceed the initial purchase price. Halons have also been identified as contributing to the depletion of the ozone layer in the atmosphere. Regulations governing the use of halons are likely to become more restrictive.

HALOTRON

Halotron extinguishers (Figure 7.21) use a halon substitute agent. They are available in 1 to 15.5 pound (0.45 to 7 kilo) units. They have a range of between 6 and 18 feet (1.8 to 5.5 meters), and a discharge time of 9 to 14 seconds. Units up to 5 pounds (2.3 kilos) are rated for class B and C fires only; units over 10 pounds (4.5 kilos) are also rated for class A fires.

FE-36

FE-36 is another halon substitute agent. Extinguishers using FE-36 (Figure 7.22) are available in sizes ranging from 4.75 to 14 pounds (2.2 to 6.4 kilos). They have a range of between 10 and 16 feet (3 and 4.9 meters) and discharge times from 8 to 15 seconds.

FIGURE 7.21 Halotron extinguisher.
(Photo courtesy of Amerex
Corporation.)

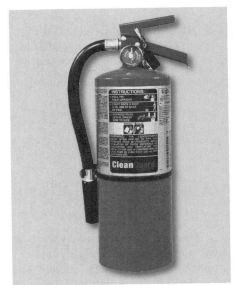

FIGURE 7.22 FE-36 extinguisher.
(Photo courtesy of Ansul
Incorporated.)

DRY POWDER EXTINGUISHERS

Dry powder extinguishers are available in stored pressure (Figure 7.23) and cartridge operated units (Figure 7.24) containing one of several available dry powder agents. This type of extinguisher is available in 30-pound (13.6 kilo) cartridge operated units. The agent is also obtainable for manual application with a scoop. Discharge times range from 30 to 60 seconds, and the discharge range is from 6 to 8 feet (1.8 to 2.4 meters). Dry powder extinguishers are also available in 150 and 350 pound (68 and 159 kilo) wheeled units (Figure 7.25).

Dry powder extinguishers are suitable only for class D fires and are the only effective fire control agents for this type of fire. Dry powder agents must be evaluated in relation to the specific combustible metal that is being protected. These extinguishers can be recharged in-house.

WET CHEMICAL EXTINGUISHERS

Wet chemical extinguishers (Figure 7.26) are designed for use on class K cooking oil fires. This is a relatively minor exposure in most industrial settings. Food preparation areas for industrial cafeterias would require this type of coverage if frying and other

FIGURE 7.23 Dry powder extinguisher, stored pressure. (Photo courtesy of Amerex Corporation.)

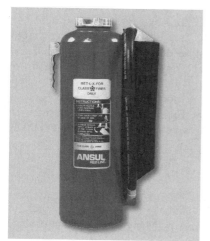

FIGURE 7.24 Dry powder extinguisher, cartridge operated. (Photo courtesy of Ansul Incorporated.)

FIGURE 7.25 Dry powder wheeled extinguisher. (Photo courtesy of Amerex Corporation.)

FIGURE 7.26 Wet chemical extinguisher (Photo courtesy of Amerex Corporation.)

oil-based cooking is done. This will also be an issue on a larger scale if the processes of the manufacturing facility involve cooking.

EXTINGUISHER RATINGS

Extinguisher ratings are used to determine which classes of fire can be controlled effectively and safely with the extinguisher being tested. Ratings also provide a guide to the size of fire an extinguisher will control. All extinguishers receive a class rating. Extinguishers for class A fires are given a numerical rating from 1A to 40A which is based on a relative scale. A 4A extinguisher, for example, will control a fire approximately twice as large as a 2A extinguisher. Class B extinguishers are rated numerically from 1B to 640B based on the approximate square footage of a spill fire they can control. A 10B extinguisher will extinguish a spill fire of approximately 10 square feet (0.9 square meters).

Fire tests are conducted using wood and excelsior for class A extinguishers. Class B extinguishers are tested on *n*-heptane fires in square pans using a liquid depth of two inches. Extinguishers for class C fires do not undergo fire tests but are tested for nonconductivity of electricity. Special tests using the specific combustible metals are employed for class D extinguishers.

The two largest testing laboratories for portable fire extinguishers are Underwriters' Laboratories and Factory Mutual. Extinguishers for use in industrial environments should have been approved by at least one and preferably both of these organizations.

EXTINGUISHER REGULATIONS

The most commonly used regulation is the National Fire Protection Association (NFPA) #10 Standard for Portable Fire Extinguishers. Many local jurisdictions have also adopted the NFPA standard as a local ordinance.

The Occupational Safety and Health Administration (OSHA) does not require portable fire extinguishers in facilities. OSHA regulations are aimed at the protection of people, not property. If extinguishers are present, OSHA regulations address the issues of placement, inspection, maintenance, and testing. If the policy of the organization allows for employee use of the extinguisher, then the regulations require employee training. The requirements of NFPA #10 were used as a basis for the initial OSHA regulations.

The insurance carrier for a specific facility may also impose requirements relative to the number, type, and placement of portable fire extinguishers. In addition, local and state authorities may have regulations in this area.

SELECTION OF EXTINGUISHERS

The selection of an appropriate extinguisher for a particular area depends upon several factors. Figure 7.27 illustrates general considerations that must be evaluated.

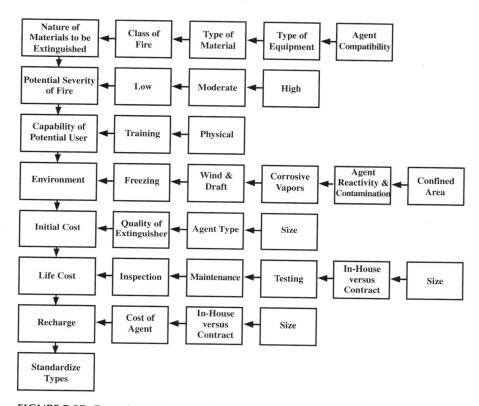

FIGURE 7.27 General considerations for selection of extinguishers flowchart.

The extinguisher must be rated for the class or classes of fire that could be expected in the area being protected. This is the first consideration when selecting an extinguisher. The process of identifying the general type of extinguisher suitable for the various fire classification combinations is illustrated in Figure 7.28. This chart helps to narrow the selection options based on the class of fire.

The materials that are being protected influence the choice of extinguisher. In a warehouse storing paper goods, for example, a water-type extinguisher would be more suitable than a multipurpose dry chemical. Although the dry chemical extinguisher is rated for class A fires, it does not offer the penetration ability that water does for deep-seated fires in class A materials.

The type of equipment and property being protected is another important consideration. Some extinguishing agents offer more rapid fire control than others but may have corresponding disadvantages like the potential to cause additional damage to the equipment. For example, dry chemical agents generally offer the most rapid control of flammable liquid fires, however, if electronic equipment is present in the same area, the residue left by dry chemical agents may cause more damage than the

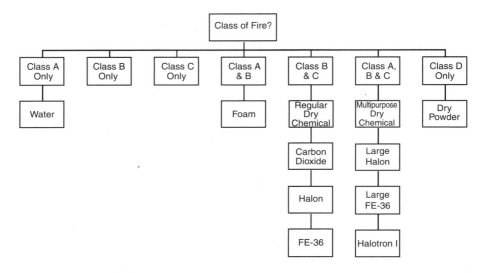

FIGURE 7.28 Type by class of fire.

fire. In this example, carbon dioxide, halon, or any of the halon substitutes would be more effective agents. The opposite situation can also occur. If an area outside the structure, a loading dock, for example, is being protected, consider that the dry chemical agent is much less affected by wind than carbon dioxide. Every extinguishing agent used for portable fire extinguishers has advantages and disadvantages that must be considered when selecting the appropriate one for protecting a specific hazard.

Agent compatibility is another consideration that will affect the choice of extinguisher. If the area being protected contains methyl ethyl ketone, a polar solvent, a regular foam agent will not be effective.

The potential severity of the fire must also be evaluated in order to determine the size of the extinguisher that will be required. If, for example, an extinguisher must be chosen to protect a quality control lab area that uses flammable liquids in quantities up to a pint, the extinguisher selected will need only limited fire control capability. If the area to be protected is a 4 by 8 foot (1.2 by 2.4 meter) dip tank, a much greater fire control capability will be necessary.

NFPA #10 divides fire severity potential into three classifications: light (low), ordinary (moderate), and extra (high). These classifications, within the context of the standard, are used to determine specific requirements for items like the distribution of extinguishers. Used in general terms during the evaluation of extinguisher protection, the hazard classifications help assess the severity of the potential fire in a specific area.

Low hazard includes areas such as offices, classrooms, and other general use areas. The quantity of class A materials in these areas is relatively insignificant.

Small amounts of class B materials may be present as well. Moderate hazard includes light manufacturing, stores, workshops, and some warehouses. The quantity of class A and B materials is greater than in low hazard areas. High hazard includes manufacturing, vehicle repair, and warehouses, where the quantity of class A and B materials is greater than in moderate hazard areas.

User capabilities also affect the choice of extinguisher. The training the individual has in using extinguishers is the primary consideration. If training will be limited or nonexistent, the extinguishers should be simple and small. Larger extinguishers can be used on larger fires. The size and capability of the extinguisher can provide a false sense of security to an untrained operator. Fighting larger fires requires more training. Occupants without sufficient training should not be engaged in fighting large fires. The untrained individual will be forced to exit the building when the agent is exhausted, and this will occur more rapidly with a small extinguisher. The user's physical ability to operate the extinguisher may be affected by physical or mental limitations. Strength limitations may also affect the size of the extinguisher that can be handled effectively by the individual.

The location where the extinguisher will be placed must be considered because the environment can have a major impact on which extinguisher will be most appropriate. Temperature will affect water and foam extinguishers. They are subject to freezing and must either not be used in unheated areas or be modified with antifreeze solutions. Wind and draft will have an impact on the effectiveness of any gaseous agent. Carbon dioxide, in particular, is difficult to use effectively in windy conditions. Corrosive vapors in the area may deteriorate the extinguisher and the materials used in its construction and may determine how easily the unit is affected by these corrosive vapors. Extinguisher valve heads are typically available in both aluminum and brass, brass being the superior choice for a corrosive environment.

Agent reactivity and contamination must also be considered because some extinguishing agents may react with certain materials and chemicals. Contamination can also occur. For example, the use of a dry chemical extinguishing agent in a food processing area would require the disposal of all exposed food items. On the other hand, a carbon dioxide extinguisher used in this same area would not contaminate the food.

The use of certain extinguishing agents in a confined area may present hazards to the individual using the extinguisher. Carbon dioxide, for example, will exclude the oxygen in an area if a sufficient concentration is allowed to accumulate. The decomposition products of halon, when heated, can also be hazardous.

The initial cost of fire extinguisher protection can vary significantly depending on what kind of extinguishers are used to achieve adequate protection. One way to improve the selection of extinguishers in terms of cost is to evaluate the facility as a whole rather than as individual areas. A decision must be made as to whether getting the most effective protection or meeting minimum requirements is the goal. Of course, the quality of the extinguishers used will have a significant impact on the cost, but this difference is often well worth the initial investment. The cost of the extinguisher should be evaluated over the life of the unit, not just based on initial purchase price. Agent prices also vary, and the agent type selected will affect the initial cost. Purple K is more expensive than regular dry chemical, but if protection is required for significant flammable liquid issues, the extra cost is justifiable.

The size selected will affect initial costs. Many small units will cost more than fewer large units of the same type of extinguisher. Placement considerations may make additional smaller units the better choice. This must be evaluated in terms of the overall protection of the facility.

When making the initial purchase of extinguishers, quotes should be obtained from several sources. Final purchase prices can vary considerably from dealer to dealer. Always buy from a reputable source, preferably one that will also provide long-term maintenance and recharging services.

Ongoing extinguisher costs should also be evaluated. For example, regular inspections will have to be performed. The time it takes to perform these routine inspections will, of course, depend on the number of extinguishers and the complexity of the inspection. The maintenance cost of extinguishers is influenced by the same factors as inspection costs. Hydrostatic testing must be performed periodically, and the costs of this maintenance and testing depend on the type and quantity of extinguishers that were selected.

Some varieties of extinguishers are easier to maintain on an in-house basis. If in-house maintenance is to be used, the selection of extinguishers can determine the cost involved and the difficulty of ongoing maintenance.

The cost of recharging extinguishers depends on the type of agent and on whether it is done in-house or by an outside contractor. How recharging of the extinguishers will be handled should be part of the initial selection criteria.

Standardization of extinguishers throughout the facility offers some advantages with respect to inspection, maintenance, and training. If extinguishers are already in place, continuing to use the same type may be the most effective selection.

PLACEMENT OF EXTINGUISHERS

Extinguishers must be placed to provide effective fire protection and to meet code requirements. Two primary criteria affect placement: travel distance, and coverage. The travel distance is the distance that an individual must traverse to get an extinguisher. Coverage refers to the number of extinguishers needed to provide adequate protection for an area based on its square footage. Figure 7.29 is a chart illustrating the maximum travel distance and square footage of coverage for extinguishers based on the hazard classification of an area. The most limiting characteristic, square footage of coverage or travel distance, must be used to place extinguishers. The illustration in Figure 7.30 demonstrates that the full area of the 75-foot (22.8 meter) circle allowed by travel distance requirements for class A extinguishers covers more than the 11,250 square feet (1045 square meters) which is the maximum allowed. The overlapping circles in Figure 7.31 demonstrate the origin of the 11,250 square feet (1045 square meters) of allowed coverage.

Extinguisher Rating	Coverage/Extinguisher (sq. ft.) Hazard Level			Maximum Travel Distance
	Low	Moderate	High	
2A	6000	3000	–	75'
3A	9000	4500	–	75'
4A	11,250	6000	4000	75'
6A	11,250	9000	6000	75'
10A	11,250	11,250	10,000	75'
20A	11,250	11,250	11,250	75'
30A	11,250	11,250	11,250	75'
40A	11,250	11,250	11,250	75'
5B	X			30'
10B	X			50'
10B		X		30'
20B		X		50'
40B			X	30'
80B			X	50'

FIGURE 7.29 Travel distance and coverage chart.

In addition to placing fire extinguishers to meet code requirements, they should be set up in areas where they will provide the most effective use. Locating extinguishers near exits (Figure 7.32) encourages the user to maintain an escape path while using the extinguisher. Extinguishers should also be placed in close, but not immediate, proximity to high hazard processes and equipment. As an example, extinguishers will be placed for the facility illustrated in Figure 7.33. To make the example less complicated, only multipurpose dry chemical extinguishers will be placed in the facility. These extinguishers are approved for class A, B, and C fires, and the numerical ratings used are 2A and 20B.

The first step is to determine the hazard classifications of the various areas of the facility. The office area is low hazard raw materials, and production and warehouse areas are moderate hazard. These hazard classes will determine the acceptable area of coverage for class A and the maximum travel distance for class B protection. The equipment and other fixed structural features must be considered when calculating travel distance to extinguishers.

The locations where the extinguishers will provide the minimum level of protection are illustrated in Figure 7.34. Usually, several placement methods will meet minimum code requirements. The various options need to be evaluated for each facility to determine which one provides the most effective fire protection. In practice, it is frequently necessary to exceed minimum standards to provide effective protection.

Another example of the office area of an industrial facility is shown both without (Figure 7.35) and then with extinguishers (Figure 7.36).

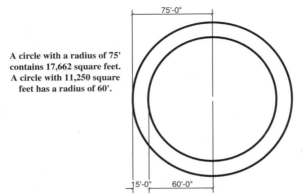

FIGURE 7.30 Travel distance vs. coverage.

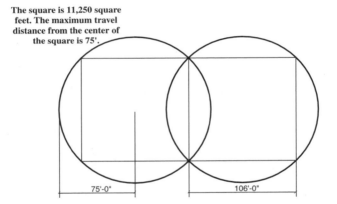

FIGURE 7.31 Travel distance vs. coverage.

FIGURE 7.32 Extinguisher mounted near exit.

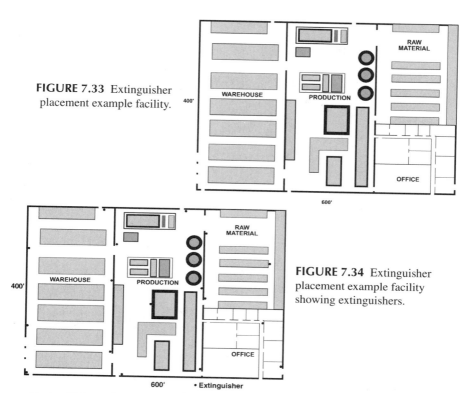

FIGURE 7.33 Extinguisher placement example facility.

FIGURE 7.34 Extinguisher placement example facility showing extinguishers.

Extinguishers should be placed in such a way that they are readily accessible and easily visible. Signs can be used to improve the visibility of an extinguisher location. All extinguishers should be securely installed on a wall (Figure 7.37) or column (Figure 7.38) or in an approved extinguisher cabinet (Figure 7.39). Extinguishers that need to be mounted outside should be provided with protective covers (Figure 7.40). According to NFPA #10 and OSHA regulations, extinguishers weighing 40 pounds (18 kilos) or less should be mounted with the top of the extinguisher no more than five feet from the floor. Extinguishers weighing more than 40 pounds (18 kilos) should be mounted with the top of the extinguisher no more than 3.5 feet (1 meter) above the floor. Both types should have a minimum clearance of 4 inches (10 cm) from the floor. Specific code requirements may apply to the installation of extinguishers and should be checked prior to setting them up in the facility.

USE OF EXTINGUISHERS

GENERAL GUIDELINES

The following general guidelines apply to extinguisher use in all situations and with all types of extinguishers. Figure 7.41 illustrates this process.

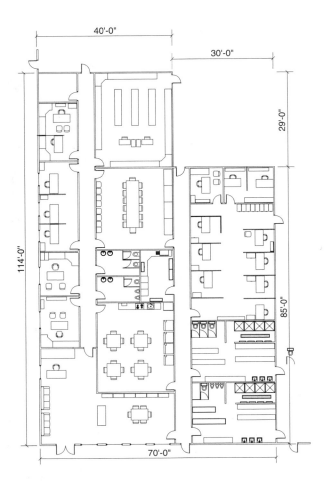

FIGURE 7.35 Extinguisher placement example 2.

When an individual discovers a fire, his/her first priority is to notify other occupants. Life safety is the primary concern. The second priority is to call the fire department, since a fire is more easily controlled in the early stages of development. Delay in summoning the fire department can make a substantial difference in the loss caused by the fire. From the fire department's perspective, it is much better to arrive and find that a fire has been extinguished with a fire extinguisher than to arrive at a fire that has grown and spread because the fire was not reported until after an occupant had unsuccessfully attempted to extinguish it. Even if the fire has been extinguished, it is a good practice to have fire department personnel check to ensure that the fire is in fact completely out.

Only after completing these first two actions should an individual begin to consider using an extinguisher. In reality, some people will attempt to fight the fire first. This is a natural human tendency that is unlikely to be changed completely. In some

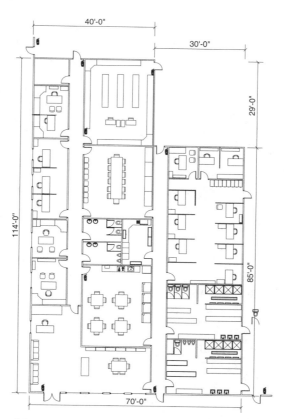

FIGURE 7.36 Extinguisher placement example 2 showing extinguishers.

FIGURE 7.37 Wall-mounted extinguisher.

FIGURE 7.38 Column-mounted extinguisher.

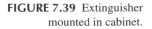

FIGURE 7.39 Extinguisher
mounted in cabinet.

FIGURE 7.40 Extinguisher
mounted outside.

cases this may be acceptable. For example, a person who has had training in the use of fire extinguishers may, depending on the size and nature of the fire, be justified in making an initial attempt to control the fire. However, this action is very risky and should be discouraged.

The ideal environment is one in which these actions can take place simultaneously. When employees work in groups, one employee can begin to control the fire while others activate the building alarm and call the fire department.

The individual must evaluate the fire situation and decide whether to fight the fire or not. This decision is based on several factors. The availability of an operable fire extinguisher is the first issue. If no fire extinguisher is accessible or one that is present cannot be operated, the rest of the considerations are academic. The next consideration is the magnitude and intensity of the fire and the rate at which the fire is growing. An individual should not risk personal injury by attempting to control the fire. If the fire has already grown too large or is developing rapidly, it is not safe to fight the fire. The person must also be concerned with the fire products being created. Depending on the nature of the fire, the smoke, heat, and fire gases generated by it can make the surrounding area extremely dangerous in a relatively short period of time.

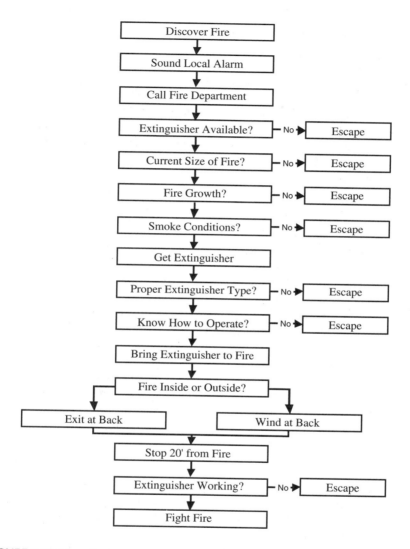

FIGURE 7.41 Flowchart on use.

If the individual determines that it is safe to attempt to control the fire, the next step is to get an appropriate fire extinguisher and bring it to the area. The extinguisher must be suitable for the class of fire to be controlled. Also, when bringing the extinguisher to the fire area, safety is more important than speed.

For any fire inside a structure, the person using the extinguisher should position himself so that an exit is always at his back. If the attempt to control the fire fails, the exit allows for a safe escape. If the fire is outside, the wind should be coming from behind the individual. This prevents smoke and heat from being blown toward the person and aids in controlling the fire by using the wind to help carry the extinguishing agent over the fire.

The person with the extinguisher must prepare to fight the fire from a comfortable distance, which should be a minimum of 10 feet (3 meters) away. Prior to moving in to begin fighting the fire, the individual must check the condition of the fire extinguisher, pull the safety pin, and test discharge the unit to ensure that it operates. Only after all of these items have been completed should the individual begin to move in to control the fire. After the fire is extinguished, the individual should always back away from the scene so that if re-ignition occurs, the person fighting the fire will see it immediately.

WATER

The pressurized water extinguisher can be used on any class A fire. The techniques used and the effectiveness of the extinguisher depend on the type of materials involved in the fire and the size of the fire.

For contained fires in trash or other receptacles, direct the discharge of the extinguisher at the inside container wall for initial knockdown of the fire. As it becomes safe to approach the container, use the thumb or forefinger to create a spray discharge from the extinguisher and thoroughly soak the materials in the container. A tool should be utilized to separate the materials to ascertain that the fire is completely extinguished.

For fires in open areas, aim the discharge at the lowest and closest edge of the fire. Sweep the nozzle from side to side, gradually working forward onto the fire. For larger areas, it may be necessary to move around the fire to reach all sections. Break up the combustibles to ensure that all smoldering materials have been extinguished.

FOAM

Both types of foam extinguishers operate like a pressurized water extinguisher, and, on a class A fire, they should be used like one. The foam extinguisher, however, is best suited for class B fires and for spills that have not ignited. The foam is discharged onto the surface of the liquid to create a blanket of foam. Foam must be applied in a way that creates minimal surface agitation. Three methods of foam application are effective.

1. Directing the discharge stream in an arch allows the foam to fall gently on the surface of the spill.
2. Discharging the foam stream against obstructions in the area and allowing it to splash and run onto the surface of the liquid is effective.
3. Aiming the discharge directly on the ground or the floor in front of the liquid allows the foam to spray and roll onto the liquid surface.

A slow, steady sweeping action of the nozzle should be used with any of these application techniques. The foam blanket must be maintained over the entire surface of the liquid. Reapplication of foam may be necessary to maintain the blanket.

DRY CHEMICAL

Dry chemical extinguishers provide the most rapid knockdown for flammable liquid fires (class B). Regular dry chemical can also be used on class C fires. Multipurpose dry chemical may be used on class A and C fires.

Stored pressure extinguishers are operated by pulling the safety pin at the head of the extinguisher, after which the hose is removed from the retaining clip. The extinguisher is then ready to discharge the agent.

Cartridge operated extinguishers are used by pulling the safety pin at the top of the cartridge (not all extinguishers of this type are equipped with this pin). Remove the hose from the retaining clip and pull it out, breaking the seal at the top of the cartridge puncture lever. Press the cartridge puncture lever while directing the top of the extinguisher away from people. This will pressurize the extinguisher. The discharge valve is at the end of the hose.

On spill fires, the dry chemical agent should be applied in a sweeping motion starting at the nearest edge of the fire area. Sweep the nozzle fairly rapidly across the fire front, extending slightly past each outside edge. Move in as the fire is controlled. The operator should never enter the spilled material. If the rear of the fire cannot be reached from the front of the spill, move along the side of the spill to reach the rear.

Fires involving an obstruction can often be fought more effectively with two people. Each individual is responsible for slightly more than half of the fire area. The basic technique is the same, but coordination between the operators is needed. Start at the center of the leading edge of the fire area. Move in and around the obstruction. Complete final extinguishment around the obstruction.

For flammable liquid in-depth fires, the basic technique is the same. However, extra caution must be used to avoid agitating the flammable liquid surface. If the agent is discharged directly into the liquid, it will spray up and out of the container and intensify the fire. The dry chemical stream must be directed such that the agent skims across the surface of the liquid.

Three-dimensional fires present a greater challenge. This type of fire occurs when the source of flammable liquid is above the floor or ground level. The fire runs down from the source to the ground or floor, creating a spill fire at the lower level. Fire fighting should begin at the lowest point of a fire. This kind of spill fire can be fought in the same manner already described. When preparing to advance the agent stream up toward the source, the sweeping action of the nozzle should be discontinued. The agent stream should be slowly moved up along the flowing flammable liquid. When the high point is reached, a sweeping action may again be needed depending on the size of the upper fire.

Pressure fires involving flammable liquids can be divided into two primary types: high sources flowing down and low sources flowing up. In all leak fires, the first action should be to shut off the flow, if possible. With liquid fires, if the leak cannot be stopped, extinguishers can be used. If the source of the leak is high and a discharge of liquid is flowing down, the agent should be directed at the source of the leak in a steady stream. Then the agent stream is slowly moved down the flowing liquid. When the ground or floor is reached, a sweeping action is slowly started and progressively wider sweeps are made as the fire is extinguished. If the sweeps are made too large too quickly, the fire will get back to the flowing flammable liquid and the process will have to be started again.

If the leak is low and discharging up, the agent should be discharged in a steady stream at the source of the leak. Gradually start a sweeping action with the nozzle, slowly increasing the width of the area covered by each sweep until the entire spill area is covered. This type of fire can be controlled most effectively with two extinguishers. One operator concentrates on the flowing liquid and the other on the spill.

Gas fires can also be controlled with dry chemical extinguishers. However, unless a fire must be extinguished to make a rescue or to enable valve shutoff, gas leaks should be allowed to burn. The risk of a gas explosion is great if the flow of gas cannot be stopped after the fire has been extinguished. Controlling this type of fire is easiest and most effective if done with a valve. If the fire must be fought, aim the stream of the extinguisher into the flow of the fire and discharge in a steady stream. The forces of the flowing gas will entrain the agent and extinguish the fire. If the source of the leak cannot be reached, a figure-eight motion with the nozzle will create a cloud of dry chemical agent.

CARBON DIOXIDE

Carbon dioxide extinguishers are most effective on electrical and small flammable liquid fires. To operate the extinguisher, pull the safety pin at the head of the extinguisher. Remove the discharge horn from the retaining clip on larger units. Smaller units have a fixed horn that must be tilted up.

An electrical panel fire can be fought by discharging the agent with side to side and up and down motions of the nozzle, thereby creating a blanket of carbon dioxide over the entire area. A small flammable liquid fire, such as on a workbench, is fought by using a sweeping action.

HALON AND HALON REPLACEMENTS

Halon and halon replacement agent extinguishers are excellent for fires in delicate equipment. They are similar in appearance and operational features to stored pressure dry chemical extinguishers. The safety pin at the head of the extinguisher is pulled first. Then the hose is removed from the retaining clip, and the extinguisher is ready for use.

When fires occur within equipment enclosures, discharge the agent into the enclosure, flooding the interior areas. Fires in unconfined areas, such as a flammable liquid spill, require a similar technique to dry chemical use. The nozzle should be swept from side to side more rapidly with halon, and a figure-eight pattern may also be used.

DRY POWDER

Dry powder is available in pails or in extinguishers. The key to effective fire control with either form is gentle application of the agent and complete coverage of the burning material.

Cartridge operated dry powder extinguishers are basically the same as dry chemical cartridge extinguishers. The same techniques are used to prepare to discharge the agent. The range of dry powder extinguishers is very short, so a close approach must be made. The agent should be gently discharged over the surface of the burning material until a layer of agent is built up over the entire surface.

WHEELED EXTINGUISHERS

Wheeled extinguishers offer greater fire control capability because of a larger agent supply. The basic concepts of use described previously for each type of fire still apply. Special training will need to be provided to individuals expected to use this type of extinguisher.

PERSONNEL TRAINING IN THE USE OF EXTINGUISHERS

All personnel should be trained in the use of portable fire extinguishers. Individuals that have received hands-on training in the effective use of portable fire extinguishers can extinguish 2 to 2.5 times as large a fire as an untrained user. Personal safety is also improved. Using a fire extinguisher without proper training may lead the user to act in an unsafe manner, risking injury or death.

The first few minutes of a fire are critical. Most fires can be controlled easily if action is taken promptly while the fire is small. Knowing how to use fire extinguishers enables employees to take the correct action early in the development stages of the fire, which can prevent large property losses.

If employees are permitted to use fire extinguishers in the workplace, OSHA regulations require that some training be provided. All employees should receive training annually. Ideally, training should include classroom instruction and hands-on practice with fire extinguishers. In low-hazard operations, however, it may be sufficient to conduct hands-on training and classroom education in alternating years. This can significantly reduce the cost of training while still refreshing information annually. If different types of extinguishers are provided, employees should receive hands-on training and an opportunity to practice with each type. Training should include information on the types and locations of fire extinguishers, principles of use, limitations of extinguishers, safety precautions during use, and response procedures to fire emergencies. Training may be conducted by in-house personnel or be handled by an outside organization such as a fire department, fire school, or consultant. Training may be conducted on site or at an off-site training facility designed for this type of training. Scheduling is usually easier with on-site training. Off-site training offers the advantage of a facility designed for this type of training and, usually, more

---

Fuel
Clean and uncontaminated
Stored in safety cans

Location
At least 50 feet from any structure
Downhill and down wind from students
Downwind from structures, parker vehicles, and other facilities

Notifications
Local fire department
Local environmental regulatory agencies
Insurance carrier
Management personnel
General employees

Extinguishers
Proper type
Sufficient quantity (usually one per student)
Delivery
Storage location
Recharging arrangements
Back up extinguishers

Safety
Pre-briefing
Training area control

FIGURE 7.42 Extinguisher training checklist.

readily available equipment. Off-site training may be required if no safe location for the training is available at the user facility.

Planning is an essential element of any successful training effort. Figure 7.42 is a brief checklist for planning and conducting an effective extinguisher training program. The checklist provides a guide to the items that must be considered by most trainers when conducting this type of training. Depending on the specific operation, some items may have to be added to the checklist while others may not apply.

Mock-ups are an important element in fire training because they make the training more realistic. A considerable amount of extinguisher training is conducted using simple devices such as a drum cut in half. Although fighting this type of fire is considerably better than no hands-on training at all, it does not provide a realistic simulation of actual fire conditions that a user is likely to encounter. The more realistic the training fire, the more effective the training experience will be in preparing the user for actual fire fighting. Figures 7.43 and 7.44 are examples of fire extinguisher training.

Many innovations have occurred in recent years in using propane fueled simulators for various types of fire training. These devices are relatively expensive, but if an organization does an extensive amount of this training, they may be well worth the investment. These devices are cleaner than liquid fueled fires and have the major safety advantage of being able to be shut off immediately if necessary.

Another issue with training is the progressively stricter environmental regulations in some areas. Burning anything for fire training may be prohibited in your area. Check these requirements carefully before conducting any live fire training.

INSPECTION OF EXTINGUISHERS

Extinguishers must be inspected regularly to ensure that they are ready in the event of a fire. Inspections should be conducted monthly. This is the frequency required by OSHA and specified in the NFPA code covering portable fire extinguishers. The monthly inspection should be thorough and cover several basic items. A flowchart of this process is shown in Figure 7.45. A checklist or form helps prevent missing items that should be examined. Some organizations maintain records on tags attached to the extinguishers (Figures 7.46 and 7.47).

When purchased, each extinguisher should be assigned an individual number that will be used to identify it in the internal record system.

The location of the extinguishers should also be checked. Make sure that the extinguisher is located where it is supposed to be and that the location is still satisfac-

FIGURE 7.43 Fire extinguisher training. (Photo courtesy of Ansul Incorporated.)

FIGURE 7.44 Fire extinguisher training. (Photo courtesy of Ansul Incorporated.)

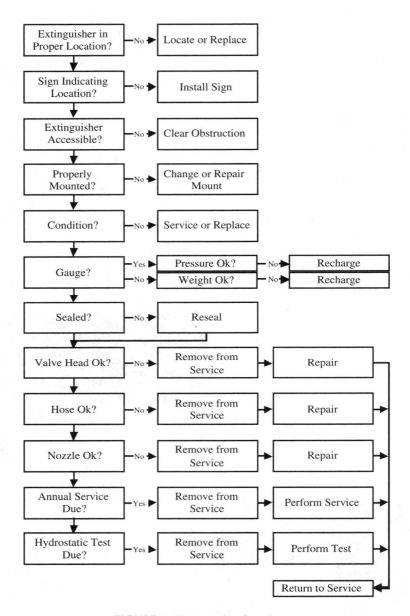

FIGURE 7.45 Inspection flowchart.

tory. Modifications to the structure, equipment position, or storage of materials may require that the extinguisher be moved to a different location. The extinguisher should be easily accessible and clearly visible. Any obstructions should be removed immediately. Any items that cannot be removed during the inspection should be noted so that the obstruction can be removed later. A note should be made even when the items blocking access to an extinguisher are removed. This allows the inspection records to reflect patterns which may possibly be corrected by relocating an extinguisher. When the extinguisher itself cannot be seen due to visual obstructions, a sign should be installed to indicate its location. The inspector should check the placement and condition of any signs.

All extinguishers should be fully charged. Extinguishers equipped with a pressure gauge can be checked visually. Extinguishers that do not have a gauge, such as carbon dioxide and cartridge-operated dry chemical extinguishers, must be weighed to check the charge. Carbon dioxide extinguisher heads are stamped to indicate a full and empty weight, as are cartridges for dry chemical extinguishers of that type. The inspector should also ensure that extinguishers are not overcharged.

Extinguishers should be sealed with a breakable seal to guard against tampering and accidental discharge. If this seal is broken and the extinguisher checks out as charged, replace the seal and make a note of the replacement.

The hose should be checked for signs of damage or wear. The tightness of the connection to the valve head of the extinguisher should also be checked. Any extinguishers with damaged hoses should be taken out of service for maintenance. The nozzle of the extinguisher should be examined for damage and obstructions. The extinguisher discharge valve should also be inspected.

FIGURE 7.46 Extinguisher tag.

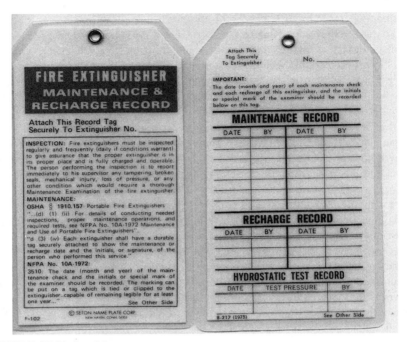

FIGURE 7.47 Extinguisher tag.

Dry chemical extinguishers are subject to caking and packing of the agent, which could prevent the extinguisher from operating properly. Caking of the agent occurs when moisture in the extinguisher causes the agent to form into clumps or a solid mass. Packing is the settling of the agent in the extinguisher to the extent that the agent forms a hard single mass. These conditions can be determined by raising and lowering the extinguisher rapidly. The inspector should be able to hear and feel a slight delay in the shifting of the agent, which indicates that it is loose.

The overall general condition of the exterior of the extinguisher shell should also be checked. It should be free of rust and corrosion, and the label of the extinguisher should be in place and legible. There should be no dents or other damage to the cylinder of the extinguisher.

One individual should be assigned the responsibility for the inspection of all extinguishers. This person may, but does not have to, be the person who actually performs the inspections. In smaller facilities, one person is usually enough. Often, an individual from the maintenance department is assigned the task of inspection. If security guards are used at the facility, the inspections can be made part of routine rounds. Outside contractors can also provide inspection services. The costs of in-house vs. contractor inspections must be compared to determine the most effective alternative. Consider all costs when making this comparison. If you contract out these inspection services, periodically check to ensure that the service being paid for is being provided. The tag shown in Figure 7.48 was discovered during an audit conducted in March; note that the monthly check for April is initialed already. This extinguisher was located on a mezzanine area of a large piece of production equipment, and further examination revealed that the contractor only visited these less accessible areas every other month.

Examples of problems discovered during inspections include obstructed access to extinguishers (Figures 7.49 and 7.50) and extinguishers not properly mounted (Figure 7.51). An extinguisher with a significantly damaged cylinder shell which was apparently struck by a vehicle is shown in Figure 7.52. Figure 7.53 shows an extinguisher with low pressure showing on the gauge and the pin and seal still in place. Figure 7.54 shows an extinguisher with the pin missing. An extinguisher where the pin was crimped over is shown in Figure 7.55. This damage would have prevented someone from being able to pull the pin.

MAINTENANCE OF EXTINGUISHERS

Fire extinguisher maintenance is essential to ensure that extinguishers will be in usable condition in the event of fire and in order to comply with regulations. Maintenance involves a more comprehensive check than the monthly inspection and is generally done once annually. Maintenance is also necessary whenever the extinguisher has been used or is found to need repair during the monthly inspection.

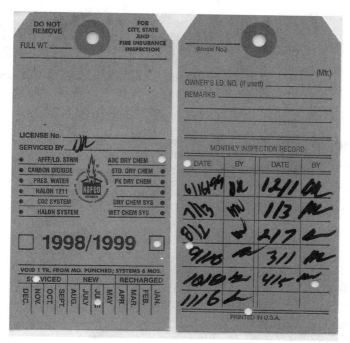

FIGURE 7.48 Extinguisher tag initialed off for an inspection not done.

FIGURE 7.49 Obstructed access to extinguisher.

FIGURE 7.50 Obstructed access to extinguisher.

FIGURE 7.51 Extinguisher not properly mounted.

FIGURE 7.52 Extinguisher cylinder damaged.

FIGURE 7.53 Gauge showing low pressure.

FIGURE 7.54 Extinguisher with missing pin.

FIGURE 7.55 Extinguisher with damaged pin.

Extinguishers require periodic internal examinations and hydrostatic tests; time intervals vary for different extinguishers. Generally, for water, foam, and carbon dioxide, the interval is five years for internal examinations and hydrostatic tests. For dry chemical, halon, and halon substitutes, the internal exam needs to be conducted every six years and the hydrostatic test every twelve years.

All organizations should keep maintenance records on extinguishers. A master extinguisher record maintained in a database is also important. Each extinguisher should be assigned a unique extinguisher number, and the manufacturer's model and serial numbers should also be recorded. The type of extinguisher should be listed, for example, "two and one-half gallon pressurized water" or "ten pound multipurpose dry chemical." Also indicate the manufacturer of the extinguisher. Maintain a record of all extinguisher purchases. The minimum information should be the supplier of the extinguisher, the date of purchase, and the initial cost. The maintenance record for the extinguisher should indicate its current location and where it is in service. The maintenance history includes the dates and reasons for any maintenance work. Who did the work and what it cost should also be recorded. When testing is performed, the name of the company as well as which type of test was performed should be noted along with the cost. The next due date should be recorded as a reminder. The use history includes all the times the extinguisher has actually been used during an emergency. A note should indicate what type of fire was fought and whether the extinguisher worked properly.

Few organizations can cost-effectively handle their own extinguisher maintenance. The necessary tools and equipment are costly. Spare parts, agents, and propellant gases must be kept in stock. Personnel must be trained initially, and they must receive periodic training to maintain their skills and stay abreast of new techniques and types of extinguishers. The organization also assumes all liability for extinguishers that are maintained in-house.

Generally, extinguisher maintenance is most effectively handled by an outside contractor specializing in extinguisher sales and service. A contractor who deals exclusively with fire protection is the best choice. Check with other facilities in the area to determine which company of this type has the best reputation for service. Clearly define what type of service will be expected before entering into any contract.

8   Emergency Planning

CHAPTER OBJECTIVES

You will be able to identify and explain:

- Types of planning
- Systematic approach to the planning process

You will also be able to:

- Develop emergency plans
- Use emergency plans

WHY IS PLANNING IMPORTANT?

Planning is an important element in the effective management of any operation. Without plans, our actions and activities would tend to occur in a haphazard manner. Planning helps to focus efforts on meeting goals and objectives with a minimum amount of resources and time. The factors that make planning useful for the management of any project are particularly crucial in the management and control of emergencies. Emergencies are, by definition, unplanned events, but this does not mean that plans cannot be prepared for handling them. The fact that emergencies are unplanned events makes it even more critical that plans be developed to identify control strategies. Emergencies can place demands on systems and personnel that are unusual and well above the commitments normally necessary. Planning can help ensure that appropriate equipment, training, and preparations to handle the foreseeable emergencies are in place.

Foreseeable emergencies should be considered, especially those with some reasonable likelihood of occurring. Some possible emergencies simply are not practical to prepare for. The costs of preparation would far outweigh the benefits.

Fires are the primary focus of this book, but there are many non-fire emergencies that may have an impact on fire protection in the facility, for example, water supply interruption due to a water main break, power interruptions that may compromise the function of installed fire protection, communications interruptions that may render the fire alarm inoperable, and severe weather or other natural disasters.

WHO SHOULD BE INVOLVED?

The personnel involved in the planning process will vary significantly from facility to facility. There are, however, several individuals that should be involved in the planning process for emergencies at any facility.

The person in charge of loss prevention and control must be involved in this process. If the facility has a fire brigade or emergency team, a representative from that group should be included in the planning process. At some phase, the plant manager or general manager of the facility should be involved. Key people in areas that may closely relate to emergency operations should be included, for example, the maintenance manager, the security chief, and supervisors from areas that are particularly likely to have fires.

Input should be received from local emergency response organizations such as fire departments, emergency medical services, and law enforcement organizations.

The balance between broad-based input and the size of the group is a critical item to keep in mind when putting the planning group together. Most plans benefit from as much input as possible, but granting membership on the planning group to anyone who has any input would create a large committee. The planning group should be small enough to accomplish the planning task with minimal delay. The best way to maintain this balance in practice is to have a small planning group which periodically holds open meetings that include all interested individuals. These meetings allow for input from all sources, while the small core group permits a more workable operational setup.

WHEN DO PLANS NEED TO BE MADE?

Plans should be prepared at several occasions during the operation of a facility. Ideally, emergency planning should begin prior to the construction of a facility. Items identified during the planning and design phase of facilities or processes can be used to develop emergency handling procedures prior to commencement of operations in the facility. Planning prior to construction also allows preventive measures to be designed into the facility. Plans should be made for the overall facility, for individual structures within the facility, and for high-hazard processes.

WHEN DO PLANS NEED TO BE CHANGED?

Any emergency planning can only be useful if the plans are current at the time of an emergency. Facilities, hazards, personnel, and many other items change on a regular basis. Plans must reflect these changes or they will become obsolete. All plans should be thoroughly reviewed and updated at least once each year. Also, any time significant changes occur in the organization which may affect the plan, the plan should be reviewed and adjusted.

HOW TO DEVELOP PLANS

A thorough analysis of the operations, hazards, and risks in the facility should be conducted. This analysis will be the basis of all planning and should be the first step in the actual development of plans. Plans are developed around the needs that exist, so an understanding of the needs is essential. Plans will also deal with the allocation of resources. A thorough analysis of the available resources is necessary to building a useful plan.

INPUT FROM OTHER PLANS

Several other plans that your organization may have prepared can be useful in the emergency planning process. OSHA-required fire prevention plans and emergency action plans can provide much useful information that can be incorporated into emergency planning. Fire prevention plans are a regulatory requirement and were discussed under hazard control. These plans deal with fire prevention issues, which are not the main focus of this chapter. Emergency action plans are required by OSHA regulations and focus primarily on emergency evacuation of employees. These plans were covered in the life safety chapter. Regulatory requirements may also have mandated the development of plans concerning hazardous materials stored or used on site. These plans may provide useful input.

SYSTEMATIC EMERGENCY PLANNING

The flow chart illustrated in Figure 8.1 shows the concept for a systematic approach to emergency planning. The first step of this process is to obtain data on the operation and facility. This information-collection phase should be very broad and general in scope. The point is to learn as much as possible about the particular operation and facility being planned. Decisions about what is and is not important are not made at this stage of the planning process. Information is needed on hazards, processes, personnel, response capabilities, and many other items.

In the second phase of the planning process, raw data is organized into useful information. Information that is not useful should be eliminated, and areas requiring additional information should be identified. The first decision is whether sufficient information has been obtained to prepare the emergency plan. If sufficient information has been obtained, the planner should move to the next phase of the process. If additional information is required, the planner go back up to the top of the process. Development of the action plan involves taking the information that has been collected and putting it into a form that will be useful at the time of an emergency.

After the initial plan has been developed, it should be reviewed to discover missing information or items which will be impractical or unworkable during an emergency. The next decision is whether the plan appears to be acceptable. If it is not, the planner should return to the action plan development stage. If the plan is acceptable, the planner should move on to the testing of the plan.

Testing emergency plans is an essential element of the planning process. Regardless of how thoroughly thought-out and prepared an emergency plan is, the only way to determine if it will work effectively during an emergency is to conduct a test. Even with actual testing, use at a real emergency can occasionally indicate problems that were not discovered during testing. This situation is exacerbated if no

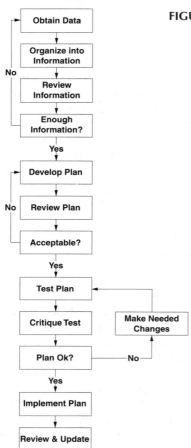

FIGURE 8.1 Flowchart on pre-planning system.

testing has been done at all. During the testing of emergency-response plans, it must be made clear to all parties participating in the test that what is under review is the plan, not individual performance. There are times and places for that individual performance review, but plan-testing is not the best time for it.

The next step in the process is to critique the test. This critique should take place as soon as possible after the actual test. Input should be received from all parties involved in the test. Any suggestions for improvement of the plan or improvement of general procedures should be taken and reviewed.

The final decision is whether the plan is acceptable or not. If the plan did not function effectively during testing, the necessary corrections and changes to the plan need to be made and retested. If the plan performed adequately or if problems require only minor revisions, these changes can then be made in the plan and the plan can then be implemented without further testing.

TACTICAL SURVEYS

Tactical surveys are performed to collect information that will be used in the planning process. Tactical surveys should not be confused with inspections. The goal of a tactical survey is to collect information that will be useful during the control of an emergency. The purpose of these surveys is not to get involved in fire prevention, recommended practices, or code enforcement. Code enforcement and fire prevention inspections are designed to institute changes that will make an area safer. Tactical surveys are designed to develop an accurate assessment of how things actually are in current practice. This comprehensive view of the advantages and disadvantages of the facility as they relate to emergency control is an essential first step.

The initial tactical survey should cover the entire facility. Although a form for use during a tactical survey is not essential, it does provide a reminder of some of the key points to observe during the survey. Figure 8.2 illustrates one example of a tactical survey form. All items that may have an impact on emergency response and control should be identified in the tactical survey. The top portion of the form provides a place for general identifying information. Obstructions to building access or areas of the building should be noted. Exposures refer to neighboring facilities that may be threatened by an emergency. Construction information about the building and specific features such as walls, fire walls, doors, and the roof should be recorded. Protection or the lack of protection for horizontal and vertical openings within the structure should be noted. Controls for utilities such as electric and gas should be noted. Information about the size of the facility should be recorded. A list of favorable and negative features about the structure should be noted. Information on fire protection systems should be included. Occupancy information such as the number of people typically in the building and whether any of these people have limitations that might prevent their escape from a fire should be recorded. Fire loading is an estimate of the fuel available to contribute to a fire and is usually expressed in pounds per square foot (kilos per square meter). This information allows for comparison with the water supply to help determine if adequate water supply is available for fire control efforts. An indication of particularly valuable items should be included. This information may allow these items to be removed prior to being damaged or lost to a fire. Hazardous materials and processes can significantly complicate fire control and pose a risk to fire fighting personnel. Knowing where these special risks are located and what type of hazards they pose should be noted. Hydrant information provides the details of nearby public water supply.

ACTION PLANS

Action plans contain the information that would be used at the time of an emergency. This information is presented in a format that makes it easy to use during emergency response. A well-prepared action plan will provide the individual managing emergency-response operations with information and guidance that will enable him/her to make more effective decisions during the emergency than would have been possible without the plan.

TACTICAL SURVEY

PROPERTY NAME	ADDRESS		
NAME	PHONE	EMERGENCY PHONE	
NAME	PHONE	EMERGENCY PHONE	
HOURS	FIRE TEAM PERSONNEL:	DAY	EVENING	NIGHT
LEVEL OF TRAINING	TYPE & AMOUNT OF EQUIPMENT		
OBSTRUCTIONS			
EXPOSURES			
CONSTRUCTION TYPE	WALLS	FIRE-WALLS	
DOORS	ROOF		
PROTECTION OF VERTICAL OPENINGS			
PROTECTION OF HORIZONTAL OPENINGS			
ELECTRICAL	GAS	HVAC	
NUMBER OF STORIES	HEIGHT	GROUND FLOOR AREA	TOTAL AREA	NUMBER OF EXITS	EXIT CAPACITY
FAVORABLE FEATURES			
NEGATIVE FEATURES			
ALARM SYSTEM			
STANDPIPE SYSTEM			
SPRINKLER SYSTEM			
SPECIAL SYSTEMS			
OCCUPANCY TYPE	NUMBER OF OCCUPANTS:	DAY	EVENING	NIGHT
ABILITY OF OCCUPANTS	FIRE LOAD		
VALUABLE ITEMS			
HAZARDOUS MATERIALS			
HAZARDOUS PROCESSES			
SPECIAL HAZARDS			

HYDRANT #	DISTANCE	GPM	PRESSURE	OUTLETS	MAIN SIZE
OTHER WATER SUPPLY					
		DATE	SIGNATURE		

(REMARKS ON BACK)

© 1981 FIRECON

FIGURE 8.2 Tactical survey form.

Action plans are designed to assist those individuals in making decisions during actual emergencies. They do not provide a script or detailed layout for specific emergency handling, but they do provide guidance that can save time during the emergency.

The scope and complexity of action plans can vary considerably depending on the hazards being planned for. Generally, brief is better. A long and detailed action plan will tend not to be used by emergency-response personnel, making the planning effort a waste of time.

It is essential that plan developers remember the primary audience of their plans. The emergency-response personnel may be reading the document in the middle of the night under extremely adverse circumstances. If the plan is not useful in that environment it will not be used.

DIAGRAMS

Diagrams are used in conjunction with plans to give the incident manager visual information. Examples might be the locations and positions of special-hazard items, fire protection safety equipment locations, and the relative positions of other items in the facility. The most effective diagrams contain the critical information that the incident manager will need during an emergency while eliminating unimportant items that maybe a distraction.

Plans should contain an overall facility diagram. At sites where several buildings are present, a diagram for each individual building should also be prepared. When individual buildings are quite large, area diagrams may also be useful. Diagrams should be drawn to scale.

It is usually best to start developing diagrams by sketching the outside walls of the building or area. A good way to get a feel for the overall layout of the facility is to start on the highest part of the roof, because this allows an unobstructed view of exterior walls. Aerial photographs can also be useful. Important building characteristics should also be shown on diagrams. Caution is necessary here to avoid a cluttered diagram. Only items and information that may have a bearing on emergency response need to be shown. If the diagram is too full of useless details, it will not be helpful during an emergency.

A legend is often the most effective way to indicate the location of items without creating clutter on the diagram. Figure 8.3 shows a sample diagram of an office building which uses numbers to indicate the location of items. These location numbers would reference back to the tactical survey form.

PLAN MANUALS

Plan manuals are consolidations of all individual plans and other useful information. These manuals are designed to allow users to access the information needed in a particular situation quickly and easily. Plan manuals should, at minimum, contain the following:

- Site diagrams
- Building diagrams
- Area diagrams
- Standard operating procedures
- Hazard information sheets
- Personnel telephone lists
- Supplier telephone lists

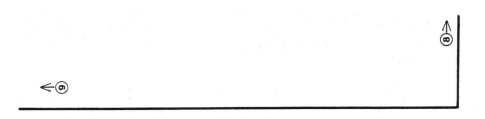
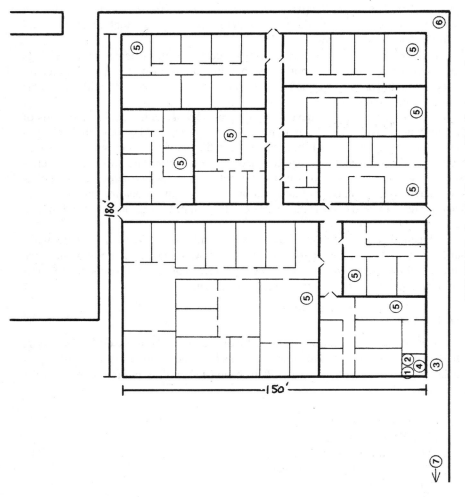

FIGURE 8.3 Sample diagram with reference numbers.

- Action plans
- Resource lists
- Evacuation plan
- Emergency notification procedures

WHERE PLANS SHOULD BE KEPT

Once planning documents are completed, it is essential that they be kept readily available for use. At least one copy of the emergency plan should be kept on file, one given to the fire department, and one maintained in an area of the facility where it can be accessed any time the employees are working. Area plans should be maintained in or near the areas they cover. Special-situation plans should be maintained in the area of the situations they cover. Additionally, copies of these sub-plans as well as a file copy should be maintained in a master location.

HOW TO USE PLANS

Plans may be used in three ways: (1) during emergencies, (2) as a training aid, and (3) to justify the acquisition and commitment of resources. That it makes sense to use plans during emergencies should be obvious, but, sometimes, in organizations with adequate emergency plans, these plans get ignored during an actual fire. The other two uses get even less attention. Part of the reason a plan may not be used during a fire is that personnel were not trained in its use. All emergency-response training provided to employees should be based on their roles and responsibilities as established in the emergency plans. Employees must be made aware that the plan exists and that certain actions are called for in the plan.

Resources are limited in all organizations. There are many competing demands for these resources. If emergency needs are to be met, every expenditure of time, effort, and funds must be justified. A well-prepared plan can provide the justification that is needed.

PLANNING FOR OUTSIDE AGENCY INVOLVEMENT

When large or complex emergencies occur, many outside agencies will be involved in the response. The success of an operation during this type of emergency is influenced by the planning that has been made for interagency cooperation and coordination.

With every increase in the number of groups responding to an emergency, the difficulties of keeping all resources focused on the incident also increase. Specific command and communications procedures need to be in place to ensure that all groups are controlled and kept informed. A breakdown in command at an incident is more likely when there are a large number of response groups at the scene.

Meetings should be conducted with representatives of each response group that may be involved during an emergency. These meetings should be held individually at first and then in a group. In the individual meetings, establish the role and relationship of the in-house responders and the way they will work with the particular outside response group. This meeting is valuable even if little substantial information is discussed. It affords everyone the opportunity to meet the representatives of other groups in a non-emergency setting.

RECOVERY PLANNING

Part of the planning process should include consideration of recovery from a fire. Assume that a fire can completely destroy your facility. What will be required to allow the business to continue operations and survive? This is the fundamental question that a recovery plan poses.

Providing for the protection of critical records is an important part of this process. All critical records must be protected both physically within the facility and by maintaining off-site duplicates. Critical records will include:

- Accounting records (i.e., accounts receivables)
- Personnel records (i.e., personnel evaluations)
- Business formation records (i.e., Articles of Incorporation)
- Customer records
- Supplier records
- Regulatory records (i.e., OSHA Injury Log)
- Product and process design information

The details of the recovery process are covered in Chapter 10.

9 Emergency Teams and Fire Brigades

CHAPTER OBJECTIVES

You will be able to identify and explain:

- OSHA requirements
- Principles of organization
- Personnel issues
- Training requirements
- Equipment needs
- Management concerns
- Emergency operations

You will also be able to:

- Determine your need for an emergency team or fire brigade
- Prepare an organizational statement
- Develop policies
- Develop an organizational chart
- Determine functions that should be the responsibility of the emergency team or fire brigade
- Select personnel
- Develop a training program
- Maintain records
- Select equipment
- Manage ongoing operations
- Manage emergency operations
- Prepare standard operating procedures

DETERMINING NEEDS

MANUAL FIRE FIGHTING OPTIONS

OSHA regulations permit individual organizations to select any of five options for manual fire fighting. These options and some of the associated requirements are illustrated in Figure 9.1.

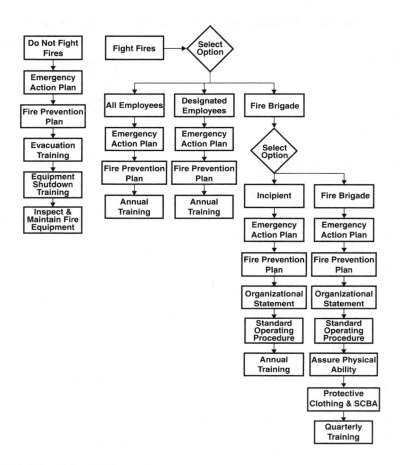

FIGURE 9.1 Manual fire fighting options.

If all employees are to evacuate the building immediately and no one fights the fire, education on the emergency action plan is required. No emergency organization is necessary. When all employees are permitted to fight fires in the facility, no organizational structure is provided. Individuals must receive education in the use of fire extinguishers and their role in the emergency action plan, but hands-on training is not necessary.

The option of assigning designated employees to fight fires results in an informal organization. The designated employees must be educated about their role during an emergency and about the proper use of fire extinguishers.

OSHA divides fire brigades into two types: incipient-stage fire fighting and interior structural fire fighting. For clarity, incipient-stage fire fighting brigades will be referred to as emergency teams, and interior structural fire fighting groups will be referred to as fire brigades. The major difference between the two is that emergency teams handle smaller, beginning-stage emergencies, while fire brigades may engage in the control of larger emergencies. Emergency teams must have organizational statements and a specific organizational structure. Members must receive annual training. Fire brigades also require organizational statements and a specific structure. All members must receive annual training, and those members that participate in interior structural fire fighting must receive quarterly training.

Throughout the remainder of this text, the term "response groups" will occasionally be used to refer to both fire brigades and emergency teams.

Which Option Is Needed?

The decision regarding which fire fighting option is most appropriate for an individual facility is based on many factors. The safety of personnel is the primary consideration. Will the use of an emergency team or a fire brigade contribute to overall employee safety without exposing the team or brigade to unacceptable risks? The availability and extent of public fire department protection must also be evaluated. In areas where quality public protection is available with a short response time, it may be difficult to justify an internal response team. However, if response times are long or the quality of protection is low, a team or brigade may be essential.

The cost of forming and maintaining an internal response organization must be analyzed and compared with the anticipated benefits. Many of the benefits, such as improved employee safety, may be hard to quantify, but adequate estimates can be made.

The contents of the building, i.e., the equipment, inventory, raw materials, processes, etc., must also be considered, because they may affect the type of fires and emergencies that can be expected. These items will also influence the capability of internal response personnel to have an impact on emergencies.

The coverage and type of installed fire protection systems and equipment will affect the need for internal response personnel. In a facility that is completely protected by automatic sprinklers, for example, the need for internal responders may be less significant. If the installed systems provide only last-resort protection, such as in areas where water damage might be extremely costly, a response team may be used to handle fires before they get large enough to activate these installed systems. Internal responders can also be used to ensure that fire protection systems are functioning properly.

The flowchart in Figure 9.2 illustrates the main points involved in the development of an in-house response capability.

EMERGENCY TEAM AND FIRE BRIGADE FUNCTIONS

If it is decided that an emergency team or fire brigade is needed, the next step should be to determine the group's responsibilities. The functions of these response groups can vary widely and must be determined based on circumstances at the specific facility. Teams and brigades generally exercise basic fire control using portable fire extinguishers, wheeled fire extinguishers, and standpipe hose lines. They should also be involved in assisting with the evacuation of personnel when needed. Internal

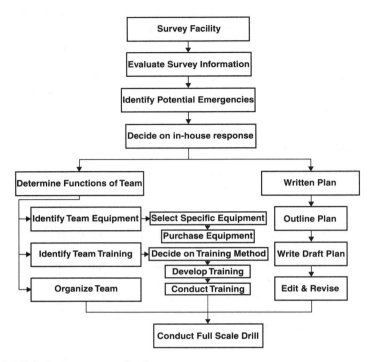

FIGURE 9.2 In-house response development.

responders provide a liaison with external response personnel. Equipment shutdown and utility and process control are tasks that can be handled more effectively by team or brigade members. The response personnel should be assigned to ensure that all fire protection features, equipment, and systems are functioning properly.

These examples include functions that are typically performed by most teams and brigades. Many other possibilities exist and will have to be evaluated based on the conditions and objectives determined at an individual facility.

ORGANIZATION

The organization of the emergency team or fire brigade will have a major impact on its effectiveness. Emergencies are fast-paced events, and without an organized response group, the emergency will outrun all efforts to control it.

ORGANIZATIONAL STATEMENT

OSHA regulations require that all emergency teams or fire brigades have a written organizational statement. The following minimum elements must be included in the statement:

1. The first section of the organizational statement should be a policy establishing the emergency team or fire brigade.
2. The organizational structure of the team or brigade must be described. This can be accomplished by a written description, organization chart, or both.
3. The type, amount, and frequency of training must also be specified. Training must be conducted at least annually for both response groups. Members of fire brigades who participate in interior structural fire fighting must receive quarterly training. It must be remembered that these are minimum requirements. The groups should receive as much training as is practical for the specific situation.
4. The organizational statement must specify the number of members on the team or brigade.
5. The specific functions which will be performed by the team or brigade must be identified.

These minimum elements must be covered by any organizational statement. The statement should also serve as a guide for the organization concerning the role and responsibilities of the response group.

Figure 9.3 is a sample organizational statement for an emergency team. Figure 9.4 is the organization chart which would be part of the organizational statement. Figures 9.5 and 9.6 are similar documents for a fire brigade.

START-UP PLANNING

If a new team or brigade is being formed, it is important to plan the start-up thoroughly. The exact purpose and function of the group should be determined. The option that is selected will have to be justified to the organization's upper management. A clearly defined purpose and function provide the best foundations for a successful justification. Management will, of course, be concerned with costs, so a thorough analysis of the costs and the expected benefits should be prepared. Benefits are sometimes difficult to quantify, but it is essential that specific figures be obtained when possible and reasonable estimates are offered in other cases.

PERSONNEL

The number and type of personnel needed for an emergency team or fire brigade will depend on the specific situation. Some basic guidelines can be used to help determine the personnel needs of a team or brigade. The number of members needed is determined by the functions that will be performed by the team or brigade. Enough members must be selected to cover all positions at all times.

A safety margin should be maintained when determining the number of people needed to perform a specific function. Allowances must also be made for vacation, sick time, and turnover. If the minimum number of personnel are assigned to the

Sample Emergency Action Team Organizational Statement

PURPOSE

The emergency action team has been established to ensure the safety of all employees, customers, visitors, contractors, and others while in the facility. This policy describes the roll of the emergency action team and outlines the responsibilities, authority, and organizational structure of the team.

SCOPE

The emergency action team will be in charge during any emergency in the facility. If an outside response organization such as the fire department, emergency medical services, or police department is called, the emergency action team will be in charge of Company personnel and will maintain a liaison with outside organizations. All Company personnel will follow the directions of the emergency action team members during an emergency.

FUNCTIONS

The emergency action team will respond to fires, smoke conditions, spills, fire alarm activations, or any other emergency within the facility. Team members will coordinate the evacuation of personnel from the building and will assist with accounting for personnel after evacuation. Members will use fire extinguishers on incipient stage fires. The team will provide advise and assistance to the fire department or any other outside response organizations.

NUMBER OF MEMBERS

The emergency action team will have 18 members on first shift and 15 members on second shift. Each member will be assigned specific responsibilities.

MEMBER QUALIFICATIONS

Members must be in good health and physically capable of performing the duties assigned to them. Each member must receive emergency action team training prior to performing duties at an emergency.

(continued)

TRAINING OF MEMBERS

Each member of the emergency action team will complete training prior to performing any duties during an emergency. Training will be provided annually for all members of the emergency action team. Training will include response procedures, use of portable fire extinguishers, stand pipe hose use, evacuation, sprinkler system and fire pump controls, evaluating emergencies, working with the fire department and safety.

TRAINING OF LEADERS

Emergency action team leaders will receive additional training of at least 2 hours annually.

ORGANIZATION

The attached organization chart illustrates the organization of the emergency action team. The team will have one team leader and two assistant team leaders. The assistants will help the team leader during emergencies and act as the team leader if the team leader is not present during an emergency. The individual assigned to sprinkler and fire pump duty will check the sprinkler system valves to ensure that they are open and confirm that the fire pump has started. The individual assigned to call the fire department and act as liaison will ensure that the fire department and any other needed outside response organizations are called. After outside response organizations arrive at the plant, they will act as liaison to these organizations. The individual assigned to electrical shut down will control power supplies as necessary. The emergency control teams are the primary and secondary groups within the emergency action team that will be responsible for emergency control. This includes the use of portable fire extinguishers, stand pipe hoses, and other activities as appropriate. The individual assigned to the evacuation area will assist with accounting for personnel and other duties as needed at the evacuation assembly area. Evacuation assistance will station themselves at strategic location within the building to assist personnel with the rapid and orderly evacuation of the building.
All emergency action team members will report to the evacuation area after completing their assigned duties.

EQUIPMENT

The emergency action team will use portable fire extinguishers and stand pipe hose.

FIGURE 9.3 Sample emergency action team organizational statement.

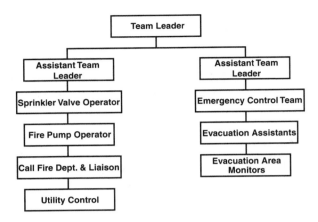

FIGURE 9.4 Emergency team organization chart.

Sample Fire Brigade Organization Statement

PURPOSE

The Fire Brigade has been established to ensure the safety of all employees, customers, visitors, contractors, and others while in the facility. This policy describes the role of the Fire Brigade and outlines the responsibilities, authority, and organizational structure of the Brigade.

SCOPE

The Fire Brigade will be in charge during any emergency in the facility. If an outside response organization such as the fire department, emergency medical services, or police department is called, the Fire Brigade will be in charge of Company personnel and will maintain a liaison with outside organizations. All Company personnel will follow the directions of the Fire Brigade Members during an emergency.

FUNCTIONS

The Fire Brigade will respond to fires, smoke conditions, hazardous materials spilled and leaks coma, confined space and industrial machinery rescue situations, fire alarm activation's, or any other emergency within the facility.

(continued)

The following functions will be performed by the Fire Brigade:
- *Coordinate and assist with the evacuation of personnel*
- *Account for personnel after evacuation*
- *Call outside response organizations as needed*
- *Provide a liaison to outside response organizations*
- *Secure the scene of emergency incidents*
- *Search for personnel who cannot be accounted for*
- *Rescue personnel from threatening fire conditions*
- *Rescue personnel from confined spaces*
- *Rescue personnel from industrial accidents*
- *Provide first aid to injured personnel*
- *Assess fire situations and decide on a course of action*
- *Use portable fire extinguishers*
- *Use wheeled fire extinguishers,*
- *Use standpipe hose*
- *Prevent and minimize damage to building and contents from fires and fire Control activities*
- *Contain hazardous material spills*
- *Stop hazardous materials leaks*
- *Other functions as needed providing such functions are within the scope of Fire brigade training*

NUMBER OF MEMBERS

The fire brigade will have 17 members on each shift. The total membership will be 52. Each member will be assigned specific responsibilities.

MEMBER QUALIFICATIONS

Members must be in good health and physically capable of performing the duties assigned to them. Prior to assignment to the fire brigade each candidate for membership will be examined by the Company physician to confirm that the individual is physical capable of performing the required duties. Each member must receive fire brigade training prior to performing any duties at an emergency.

TRAINING OF MEMBERS

Each member of the fire brigade will complete training prior to performing any duties during an emergency. Training will be provided quarterly for all members of the fire brigade. Training will include response procedures, use of portable fire

(continued)

extinguishers, use of wheeled fire extinguishers, stand pipe hose use, fire control operations, evacuation, sprinkler system and fire pump controls, hazardous materials incident handling, search and rescue, confined space rescue, industrial machinery rescue, first aid, evaluating emergencies, working with the fire department and safety.

TRAINING LEADERS

Fire Brigade leaders will receive additional training of at least 8 hours annually.

ORGANIZATION

The attached organizational chart illustrates the organization of the fire brigade. The fire brigade will have one fire brigade chief and three assistant chiefs, one for each shift. The assistants will help the chief during emergencies and act as the chief if the chief is not present during an emergency. There are four captains. One for each specialty team. There is a rescue, fire control, hazardous materials, and support team. The rescue team is responsible for assisting with an evacuation and with accounting for personnel at the evacuation assembly area. The rescue team will also search areas for individuals that are not accounted for after an evacuation and ensure that areas threatened by an emergency have been cleared of all personnel. The rescue team will perform or assist outside response personnel with confined space, industrial machinery, and other types of rescue. The fire control team will be responsible for all fire control operations. This includes the use of all types of portable and wheeled fire extinguishers, stand pipe hose lines, and special extinguishing agents. The hazardous materials control team is responsible for controlling emergencies involving hazardous materials spills and leaks. The support team will check the sprinkler system valves to ensure that they are open and confirm that the fire pump has started. This team will also call the fire department and any other outside response organizations that are needed. After outside response arrive at the plant, they will act as liaison to these organizations. The support team will control all utilities and provide salvage functions as necessary.

EQUIPMENT

The fire brigade will use: firefighting protective clothing, self-contained breathing apparatus, portable fire extinguishers, stand pipe hose, wheeled fire extinguishers, rescue tools and equipment, hazardous materials equipment, and salvage equipment. All equipment provided for fire brigade use will be inspected at least monthly.

FIGURE 9.5 Sample fire brigade organizational statement.

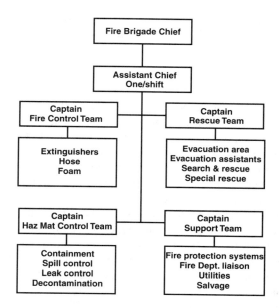

FIGURE 9.6 Fire brigade organization chart.

brigade and someone quits or is transferred after the annual training session, a new member will have to be selected and trained. If a few extra people have been assigned to the brigade, this will not be a problem. The tendency is to underestimate the number of people needed, so it is usually good practice to add a few people to the initial estimate.

For critical tasks, a backup person should be assigned to the team. Calling the fire department, checking sprinkler valves, and ensuring the fire pump is running are a few examples of critical tasks. A planner should avoid giving too many functions to one person. The three examples just given could all be done by one person, but it would take too long. At least two people should be assigned to perform these functions, or even three people may be necessary if the sprinkler valves are located away from the fire pump.

Personnel assigned to critical tasks may have secondary assignments, but no individual should be given more than one critical task. For example, the person assigned to call the fire department could also be the liaison person after the fire department arrives. The time it takes the fire department to respond should be adequate for the individual to get to the appointed meeting area after making the initial call.

The functions that the team or brigade will be expected to perform should be carefully analyzed to determine the number of members needed. Response group membership is often underestimated. It is far more effective to select and train a few extra people than to have too few.

The ideal method for getting personnel is to select volunteers who are interested in participating in the team or brigade. In some cases, however, it may be necessary to assign people to the response group.

SELECTION CRITERIA

Several selection criteria can be used to assist in getting the best personnel. All members of the team or brigade must be immediately available during an emergency. If an individual's regular job does not permit him to immediately stop what he is doing, he should not be assigned to the team. Personnel with other emergency duties are not good candidates for teams and brigades. For example, security officers should not be the only brigade members because they have other responsibilities that must be handled during an emergency. Some overlap between security and the team or brigade will improve coordination and effectiveness, but the entire security function cannot be abandoned during an emergency. Other types of members must be available.

Personnel with critical job skills or experience should be encouraged to join the team or brigade. Maintenance personnel are a good example because they know the facility, equipment, and utility system. This knowledge can be a major advantage during emergency control.

All personnel should be in good physical condition. Assuring physical capability is only required by regulation for fire brigade members who perform interior structural fire fighting, but emergency control of any type can be physically demanding and stressful. All members of the response group should be in good health and reasonable physical condition. All personnel should receive a medical checkup prior to being assigned to the team or brigade. Annual medical exams would be a good investment. A medical monitoring program must be established for members who respond to hazardous material emergencies. This monitoring program is designed to track exposure to hazardous chemicals.

A membership application can be used to acquire much of the information described here. Figure 9.7 illustrates a sample fire brigade membership application.

RECORDS

A record system should be established for response group personnel. Each individual assigned should have a record documenting the training he/she has received, the incidents he/she responded to, and his/her medical exam history. Figure 9.8 illustrates a sample individual training record form. These records may be more easily maintained in a computer database. This form is used to record the training a specific individual has received. The actual medical records should not be maintained by the brigade or team for legal reasons.

TRAINING

The training that fire brigades and emergency teams receive is one of the most important factors in the performance of these groups. All areas of operation are affected by the quality and quantity of the training provided. Training should include a mix-

FIRE BRIGADE MEMBERSHIP APPLICATION

NAME

DATE

DEPT.

SUPERVISOR

POSITION

EMPLOYEE #

SHIFT

PHONE #

What skills do you have that may benefit the fire brigade?

Why do you want to be a member of the fire brigade?

What do you believe you can do for the fire brigade?

Do you have any phobias (i.e. fear of heights) that might limit your ability to participate in fire brigade activities?

What training in emergency response have you had within the last two years?

Do you have previous emergency response experience?

What type of emergency response organization?

How long have you been an emergency responder? What is your current position?

Can you leave your regular job on short notice?

SIGNATURE

© 1989 FIRECON form FBAPP

FIGURE 9.7 Membership application form.

ture of classroom and hands-on instruction. Classroom training is necessary to build background knowledge and familiarity with concepts, while hands-on training and skill practice sessions are essential to developing ability and confidence.

The amount of training needed will vary depending on the functions of the specific fire brigade or emergency team. All fire brigades that engage in interior structural fire fighting are required to have quarterly training. Emergency teams must

INDIVIDUAL TRAINING RECORD

NAME			POSITION	
SHIFT	EMPLOYEE #		PERIOD COVERED	

DATE	HOURS	TOPIC	INSTRUCTOR

© 1989 FIRECON form INDTRGRC

FIGURE 9.8 Individual training record.

have annual training. These are the minimums established by OSHA. The amount and quality of training provided for the brigade or team will, more than any other single factor, determine the capability of the group.

All members of the team or brigade must have sufficient training to perform their duties in a safe and effective manner. This may require more frequent training than

---

Emergency Team Training

1) Introduction
   a) Instructor
   b) Objectives of training
      i) Build knowledge and skills
      ii) Improve confidence
      iii) Make response group
         operations more effective
   c) Topics
   d) Purpose of response group
      i) Life safety
      ii) Emergency containment and
         control
      iii) Property conservation
   e) Organization of the response group
   f) Response
      i) Notification
      ii) Procedures
   g) Command structure
2) Fire behavior
   a) Elements of fire
   b) Classes of fire

   c) Products of combustion
   d) Heat transfer
   e) Types of fuels
   f) Source of ignition
3) Evacuation
   a) Alerting occupants
   b) Assisting occupants
   c) Accounting for occupants
4) Installed systems
   a) Checking control valves
   b) Checking pumps
   c) Safety around special installed
      systems
5) Portable fire extinguishers
   a) Types and characteristics
   b) Selection
   c) Operation
   d) Safety
   e) Hands-on practice
6) Working with the Fire Department
   a) Roles and responsibilities
   b) Communications
   c) Command structure

FIGURE 9.9 Emergency action team model outline for basic course.

the minimums established by OSHA regulations. In addition, response group leaders and instructors must have a higher level of training than the general membership.

All team members require training in several basic areas. They should: (1) understand the role and purpose of the team or brigade, (2) for their own safety, have a basic knowledge of fire behavior and spread, (3) know how to assist in the evacuation of the facility and how to account for the personnel after an evacuation, (4) know how to report an emergency to outside response organizations and how to work with these groups after they arrive, and (5) have an understanding of fire extinguishers and the skills necessary to use them effectively.

Figure 9.9 is a sample outline for the initial training of an emergency action team. Figure 9.10 is a sample of the additional items that would need to be covered during an initial training program for a fire brigade. These samples are designed to provide guidance concerning some of the items that may need to be covered in this type of training. Effective training must be tailored to the needs of the individual group and to the specific functions identified in the response group's organizational statement. These sample outlines are basic training programs identifying only the minimum elements. The final version of an initial training program should be expanded to meet the specific needs of your facility.

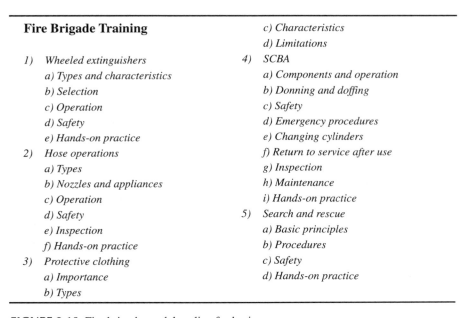

Fire Brigade Training

1) Wheeled extinguishers
   a) Types and characteristics
   b) Selection
   c) Operation
   d) Safety
   e) Hands-on practice
2) Hose operations
   a) Types
   b) Nozzles and appliances
   c) Operation
   d) Safety
   e) Inspection
   f) Hands-on practice
3) Protective clothing
   a) Importance
   b) Types

   c) Characteristics
   d) Limitations
4) SCBA
   a) Components and operation
   b) Donning and doffing
   c) Safety
   d) Emergency procedures
   e) Changing cylinders
   f) Return to service after use
   g) Inspection
   h) Maintenance
   i) Hands-on practice
5) Search and rescue
   a) Basic principles
   b) Procedures
   c) Safety
   d) Hands-on practice

FIGURE 9.10 Fire brigade model outline for basic course.

A task analysis should be performed on each function that members of the team or brigade will be expected to perform. A task analysis breaks a task into its individual components so that the specific knowledge and skills required to perform the task effectively may be identified. The distinction between knowledge and skill is important because they will need to be taught differently. Knowledge may be transmitted in many ways, while skill development requires hands-on practice.

Performance objectives are developed for training based on the information obtained during the task analysis. Performance objectives specify the desired learning outcome. They identify the knowledge or behavior (skill), the conditions under which the performance will occur, and how well the students must perform. An example of a performance objective is "the fire brigade member will don an SCBA from a carrying case, completing all steps accurately, while wearing protective clothing, in no more than 60 seconds."

All training will get improved results if it has been well planned. Lesson plans and lesson outlines are valuable aids for the instructor. These materials should be prepared in advance of the training program and used to guide the training process. They are not a substitute for a qualified instructor, but they will help a competent individual to deliver higher-quality training.

The lesson plan (Figure 9.11) provides, as the name implies, a plan for teaching the lesson. It identifies the topic covered by the lesson. Lesson plans can be prepared for each individual topic within a session or just for each separate session. The level of instruction is identified, giving the target level of competence and specifying whether to expect familiarization, mastery, or some level in between. The

Topic:	Recommended audiovisuals:
Level:	Handouts:
Performance objectives:	References:
	Presentation:
Timeframe:	Application:
Method of instruction:	Summary:
Materials required:	Evaluation:

FIGURE 9.11 Outline.

Performance Objectives section provides for the detailing of the final objectives of the instruction in this lesson. These objectives specify what the students should know and be able to do at the conclusion of the lesson. The time frame provides the instructor with an estimate of the time required to complete the lesson. The method of instruction refers to the methods, such as lecture or demonstration, which should be used to teach the lesson. The materials required are itemized. This list provides a reminder of all the materials needed to conduct the lesson, for example, flipcharts, practice equipment, or audiovisual equipment. The Recommended Audiovisual section provides a list of slides, films, videotapes, or other audiovisual materials that may be useful in teaching the lesson. A list of handouts is also provided. The Reference section lists the references that were used to create the lesson. These can be used as a source for the instructor to refresh himself on the topic or to find answers to specific questions that may be raised during training. The Presentation portion of the lesson plan refers to the lesson outline. Application refers to the things students will do to apply the information provided in the lesson. For example, if the lesson is on fire extinguisher use, the application phase should provide for hands-on use of extinguishers. The summary gives a brief review of the main points of the lesson and should also motivate the students to retain and use the information. The Evaluation section contains guidance on testing and other evaluation methods that should be used to confirm student understanding.

The lesson outline (Figure 9.12) provides a format for the information that an instructor will use during the presentation of the lesson. The AID column allows the instructor to make notes on the audiovisual or other types of aids that will be used. Teaching points are the notes that remind the instructor of the material to be covered in the lesson and also help the instructor maintain the proper sequence. The Notes column is for supplemental information.

AID	TEACHING POINTS	NOTES
	I. Main Topic Division	Reference
	A. Main Points Within Topic	
	1. Secondary points within topic	Time check
	a. Detail points	
		Additional details
SL	Slide	
VT	Video tape	
TR	Transparency	
HO	Handout or manual page reference	
CH	Chart	

FIGURE 9.12 Lesson outline.

The scheduling of training is often a challenge. Sufficient time is needed to conduct training, but time away from production is often difficult to get approved and complex to schedule. Many options are available for scheduling training, and the more flexibility built into the training schedule, the better. Management has a reasonable concern that production and normal work schedules should not be unnecessarily interrupted. This concern must be addressed to gain approval for training programs.

Each training session should be documented. A training report form (Figure 9.13) is useful for recording basic information about the training session and ensuring that attendance can be verified. Documentation of training should be maintained permanently.

When training involves live-fire practice sessions, special caution must be used to ensure that the training is safe and effective.

Students should be permitted to evaluate training sessions so that the effectiveness of the training may be improved over time. Figure 9.14 is an example of a form that may be used for this purpose.

During hands-on training it is often useful to work with a checklist for evaluating student performance. A sample checklist for self-contained breathing apparatus (SCBA) training is shown in Figure 9.15. The three check-off columns are Fail, indicating that the performance does not meet the minimum level required; NI, indicating needs improvement but meets the minimum; and Comp for completed satisfactorily.

If numerous hands-on practice activities will be performed during the training session, particularly if the group is large, a checklist for student participation (Figure

TRAINING REPORT	
Date: Subject: Sponsor: Location: Instructor:	
PRINT NAME	**SIGNATURE**
1	
2	
3	
4	
5	
6	
7	
8	
9	
10	
11	
12	
13	
14	
15	
16	
17	
18	
19	
20	
21	
22	
23	
24	

Instructor Signature _____

FIGURE 9.13 Example training report.

9.16) may be helpful. The hands-on activities may be organized into various evolutions and students checked-off as they complete each numbered skill practice area. For example, if each individual needs to practice with a stored pressure dry chemical extinguisher, a cartridge operated extinguisher, a carbon dioxide extinguisher, and a standpipe hose, each of these activities may be numbered one through four, and as students complete each task, they are checked off in the appropriate column of the form.

TRAINING EVALUATION	
Date:	
Topic:	

What portion of the training was most beneficial? Why?

What portion of the training was least beneficial? Why?

If you could change one thing about this training program, what would you change?

Item				
Course content				
Course materials				
Course organization				
How would you recommend this course to others?				
Instructor's presentation ability				
Instructor's knowledge of the subject(s)				
Instructor's handling of participant questions				
Instructor's overall effectiveness				
How would you recommend this instructor to others				
Overall course effectiveness				
Additional Comments				

FIGURE 9.14 Training evaluation form.

The effectiveness of the response group is going to reflect the effectiveness of their training. Do not skimp in this area.

EQUIPMENT

To effectively accomplish the mission of controlling emergencies, the response group must be properly equipped. Equipment requirements should develop directly from the function list prepared during the organization of the response group. Too frequently, equipment needs are considered before organization is complete. The functions should determine the equipment, not the reverse. If equipment is consid-

Skills Checklist
Donning, Use, and Doffing SCBA

#	Task	Fail	NI	Comp
1.	Get SCBA			
2.	Pre-inspection			
3.	Check pressure			
4.	Position SCBA for donning			
5.	Position self for donning			
6.	Grasp unit			
7.	Raise unit			
8.	Swing onto back			
9.	Chest strap			
10.	Shoulder straps			
11.	Position and snug unit			
12.	Waist strap			
13.	Open cylinder valve			
14.	Check face-piece			
15.	Don face-piece			
16.	Position face-piece			
17.	Tighten straps (proper order and correct technique)			
18.	Test seal			
19.	Correct seal problems (if necessary)			
20.	Test exhalation valve			
21.	Connect low pressure hose			
22.	Work while wearing unit			
23.	Shut down operation & remove face-piece			
24.	Remove unit			
25.	Close cylinder valve			
26.	Bleed pressure			
27.	Extend straps on unit			
28.	Extend straps of face-piece			
29.	Clean face-piece			
30.	Remove used cylinder			
31.	Install new cylinder			
32.	Test unit			
33.	Stow unit in case			

FIGURE 9.15 Skills checklist form.

ered the primary issue, the response group will have many items of equipment but may not truly be well equipped for their purpose. The list in Figure 9.17 provides a beginning point for considering the equipment that the response group will need. This list is not intended to be a comprehensive list of all the equipment that may be necessary.

Personal protective clothing and equipment are essential for a fire brigade and useful for the emergency team. The protective clothing ensemble for fire brigades consists minimally of a helmet, hood, coat, pants, boots, gloves, and self-contained breathing apparatus. For emergency teams, no protective clothing is required, but it is good practice to provide at least gloves and head protection.

A self-contained breathing apparatus (SCBA) is essential when emergency operations are conducted in a hostile or potentially hostile environment. The SCBA should be provided for all fire brigade members who may be exposed to dangerous atmospheres. The chart in Figure 9.18 illustrates the main portions of an SCBA program.

STUDENT CHECKLIST

DATE	PROGRAM									
	STUDENT NAME	1	2	3	4	5	6	7	8	
1										
2										
3										
4										
5										
6										
7										
8										
9										

FIGURE 9.16 Student checklist form.

Provision of an SCBA involves ensuring that an adequate number of SCBA are readily available for use by response group members. These SCBA should be stored in a secure, clean, and accessible area. Storage areas can be located strategically throughout the facility or in a central location, depending on the layout of the site. If a central storage area is used, a method for transporting the SCBA to the scene of an incident should be available. All SCBA must be certified by the National Institute of Occupational Safety and Health (NIOSH). A replacement program should be used to ensure that SCBA are periodically replaced. Using a programmed approach to replacement allows more accurate budgeting and permits the expenses of replacement to be spread over several budget years.

Portable fire extinguishers	*Self-contained breathing apparatus (SCBA)*
Wheeled fire extinguishers	*Airline breathing apparatus*
Standpipe hose	*First aid kit*
Flash lights	*Back board*
Portable lights	*Stretcher*
Radios	*1 hose, fire service quality*
Structural firefighting protective clothing	*1 nozzle, fire service quality*
Helmet	*2 hose, fire service quality*
Hood	*2 nozzle, fire service quality*
Coat	*Salvage covers*
Gloves	*Squeegees*
Pants	*Sprinkler head wedges*
Boots	*Extra sprinkler heads*
Chemical protective clothing	*Camera*
Disposable splash suits	*Pry bar*
Splash suits	*Foam*
Gloves	*Foam nozzle*
Boots	*Foam eductor*
Face shields	*High expansion foam*
Goggles	*High expansion foam generator*
Respirators	*Smoke ejector*

FIGURE 9.17 Basic equipment list.

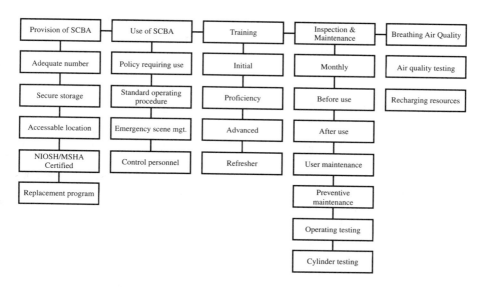

FIGURE 9.18 SCBA model program.

The use of SCBA during emergencies should be required by fire brigade policy. Standard operating procedures should establish the conditions under which the use of SCBA is required as well as other general provisions of use, for example, the practice of always working in pairs while using SCBA. This buddy system is one of the primary safety elements involved in SCBA use. Command personnel must establish a method for monitoring the use of SCBA at an emergency so that operations can be managed effectively. A system to control and account for personnel is essential. All individuals working with SCBA in a dangerous environment must be controlled.

Training in the use of SCBA is essential to effective operations. Fire brigade members should receive training on a regular basis to maintain skill and should periodically receive advanced training to improve capabilities. A certification system is the most effective method for ensuring that all personnel are competent in the use of SCBA prior to allowing any emergency scene use of SCBA. This certification should be an annual requirement.

All SCBA require regular inspection and maintenance and should be checked thoroughly each month. An example of a monthly inspection form is shown in Figure 9.19. Qualified users should perform a check before and after use. With proper training, usually from the manufacturer, members of the fire brigade can be taught to perform some basic maintenance functions. If the fire brigade has many SCBA units, this may less expensive than outside maintenance. A detailed functional check of each SCBA should be made at least once annually. Cylinders must also be tested and maintained.

Breathing-air quality is another issue involved in the SCBA program. Air quality should be tested at least once annually. If an outside contractor refills the SCBA, documentation of appropriate testing will be sufficient. Adequate resources for recharging used SCBA cylinders must be available. If recharging facilities will not be readily available during an emergency, spare cylinders should be purchased.

All items of response group equipment should be inspected at least once monthly. Adequate records of the inspection, maintenance, and use of equipment should be maintained.

MANAGING DAY-TO-DAY OPERATIONS

The routine day-to-day operations of an emergency team or fire brigade require time and effort, but this is often overlooked. If a response group is only considered during emergencies, it will not be effective.

One of the most difficult items is staffing. Effective response group operations can only occur when adequate personnel are available. Turnover, vacations, shift and job changes, sick days, and many other items can have an impact on the manning of the response group. Managing this staffing can be challenging, but it must be done effectively to ensure response group readiness. Overstaffing is probably the simplest way to address this problem. If your operation must have at least twelve members to perform adequately, assigning fifteen or eighteen members will help to ensure that enough people are available at any given time. This can, however, increase training

SELF-CONTAINED BREATHING APPARATUS INSPECTION			
UNIT # LOCATION			
ITEM	Inspector should initial blocks that do not require specifc information.		
Date of inspection?			
Is unit in proper storage location?			
Is storage location unobstructed?			
Is storage area clean?			
Is the case in satisfactory condition?			
Is the inside of the case clean?			
Is the unit clean?			
Are all components present?			
What is the cylinder pressure?			
Is the cylinder gauge in good condition?			
What cylinder number is on the unit?			
Is the cylinder in good condition?			
When was the last hydrostatic test on the cylinder?			
Is the high pressure hose connection tight?			
Is the high pressure hose in good condition?			
Does cylinder valve operate properly?			
Does the regulator function properly?			
Does the remote pressure gauge operate properly?			
Do all regulator valves function properly?			
Is the low pressure hose in good condition?			
Is the facepiece view area in good condition?			
Is the head harness in good condition?			
Are the head harness straps fully extended?			
Does the exhalation valve function properly?			
Are body harness straps in good condition and fully extended?			
Does the low air pressure alarm function properly?			
COMMENTS and CORRECTIVE ACTIONS TAKEN			
SIGNATURE OF INSPECTOR			

FIGURE 9.19 SCBA monthly inspection form.

and other costs. The brigade leadership should manage this area for optimum performance by closely tracking membership and staffing levels to establish the actual minimum manning level and identify how many members need to be assigned to ensure that this number of personnel is always available. Brigade leadership should insist that response group members keep them informed about changes in their status, ideally providing notification of future changes ahead of time so that adjustments can be made in response group manning.

Budgeting is another challenging issue. Response group leadership must sell its goals to management to gain the approvals and funding that are needed to support response group functions. Planning is an essential part of this effort. Leadership must evaluate items like equipment purchases well ahead of when the equipment is needed so that adequate time is available to sell the need to management and get funds approved. The more the response group can be run like any other aspect of the business, the more respect management will have for the operation.

NONEMERGENCY FUNCTIONS

A fire brigade or emergency team is a substantial investment. To gain the maximum benefit from this investment, these groups should be used to perform non-emergency functions in addition to emergency response. The training and organization required for effective emergency operations provides a valuable background for several related functions. At a minimum, the brigade or team should be involved in (1) fire prevention, (2) fire protection equipment and systems inspection, and (3) emergency planning. Using the response group for these non-emergency functions improves the utilization of the group in two ways. First, performing these functions helps the team or brigade stay up-to-date on the facility and emergency control equipment and systems. Second, the members of the response group are the employees most concerned about the condition of equipment, systems, and plans because they are the ones who will have to use these items during an emergency.

FIRE PREVENTION

The prevention of unwanted fires should be part of every employee's responsibilities. The members of the emergency response group have an advantage in this area due to the training provided for emergency response. The group will be more aware of the types of fire hazards. This improved awareness makes the response group an excellent fire prevention team. Members can rotate inspection responsibilities by changing areas of coverage, for example, once per quarter. In this arrangement, each brigade member is assigned an area to monitor for fire prevention. The members make regular surveys in their areas at least once per month and they make spot checks more frequently.

These surveys improve fire prevention and also keep brigade members familiar with all plant areas. This improves both prevention and emergency response.

FIRE PROTECTION EQUIPMENT AND SYSTEMS

Fire brigade members can be made responsible for routine inspection and maintenance of fire protection equipment and systems. As with fire prevention, this improves both preparedness and response. The systems are maintained more effectively, and brigade members stay familiar with the systems and equipment in the facility.

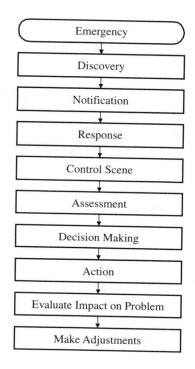

FIGURE 9.20 Emergency response process.

EMERGENCY PLANNING

All facilities should have some type of emergency plan. Response group members are ideal for helping to develop these plans and ensuring that they are kept current. The training they receive in emergency response helps the response group members identify what items need to be planned and understand how the plans will work effectively. These personnel will also play a major role in the use of plans during an emergency, so their involvement in the development of the plans establishes a vested stake in the process that will encourage those personnel to make use of the plans.

EMERGENCY OPERATIONS

EMERGENCY RESPONSE PROCESS

The flowchart in Figure 9.20 shows the major steps in the emergency response process. These steps stay relatively constant for most types of emergencies. The items that must be completed in each step may vary significantly depending on the magnitude and scope of the emergency, but the fundamental steps are fairly consistent.

The emergency must be discovered before any emergency response may begin. This may occur immediately or be delayed. Delayed discovery allows the emergency to get worse without intervention and will make the challenges of emergency response more difficult.

Notifications may occur at a few phases of the emergency. The initial notification to the in-house response team is the most important for this discussion. Response is the phase of the emergency where resources, including personnel, are brought to the scene. Controlling the scene is one of the first priorities of the response team. People must be cleared from the area for their own safety and in order to allow responders room to work effectively.

Assessment is the phase where responders collect and evaluate information concerning the emergency. Decision-making occurs based upon the assessment. Actions are then taken to implement the decisions. Additional assessment is conducted to evaluate the impact that the actions taken have had on the emergency. If the actions were appropriate and control is being achieved, then no changes are needed. Emergencies do not tend to progress this smoothly, however, and often the last step will be required. Adjustments must be made to the responders' activities based on feedback from the emergency.

GENERAL EMERGENCY RESPONSE

Standard Operating Procedures

Some of the basic elements of emergency response are the same for any type of emergency. These general emergency response components provide the foundation upon which the more detailed aspects of handling a specific situation can be built. Standard operating procedures (SOPs) are among these foundation elements. Standard operating procedures are standardized approaches to certain types of operations or to specific phases of an operation. They outline procedures that can be applied to most, if not all, situations covered by the procedure, without modification. SOPs serve as a checklist to ensure that all essential elements of a procedure are accomplished in the proper sequence. They provide a common area of understanding among various response personnel.

The standard portion of the SOP implies that a certain activity or series of activities is accomplished in the same manner from situation to situation. This consistency is beneficial at the emergency scene because it helps reduce the many variables which must be considered during the control of an emergency. The operating portion of the SOP indicates that SOPs deal with activities and tasks. These can be small portions of what occurs during emergency response or major sections of the response effort. The procedure portion of the SOP addresses the methods and techniques that are used to accomplish the purpose of the SOP. Procedures for basic activities that can be predicted allow us to concentrate on thinking about those portions of the incident response that do not fit well into established procedures.

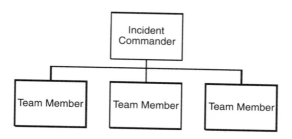

FIGURE 9.21 Incident command system, basic.

All incidents are different. You can respond to emergency incidents every day for a year and will not respond to two identical incidents. However, incidents do have similarities. Each of those 365 incidents will have certain basic characteristics in common. SOPs focus on reducing the time and effort spent considering the elements that remain the same for many different types of incidents. This allows you to concentrate on the portions of the emergency that are unique.

Incident Command System

The incident command system (ICS) provides a framework for the management of emergencies regardless of type. The incident command system is designed to provide an effective means of ensuring that all functions at an emergency are appropriately managed. Figure 9.21 illustrates the most basic organizational structure of the incident command system. Figure 9.22 illustrates the incident command system with two individual task units. The addition of unit commanders may be necessary as emergencies become larger or more complex.

The incident commander has the ultimate responsibility for the entire incident. There is only one incident commander, but this responsibility may change hands during the course of an incident. Unit commanders are responsible for the operations of a single unit of resources. These individuals report to the incident commander. Each unit is assigned a specific task. Staff officers may also be used in this system and they report to and assist the incident commander. These individuals are responsible for specific areas of the operation. The resource officer is in charge of obtaining resources needed to control the incident. This officer is also responsible for maintaining a liaison between plant response personnel and outside agency responders. The safety officer is responsible for ensuring safe operations during the incident. The medical officer is responsible for medical information, first aid treatment, and medical transportation. The public information officer should be the sole source of information to release to the news media.

The command system may become more complex than described here, but, typically, fire brigade operations do not exceed what has been described.

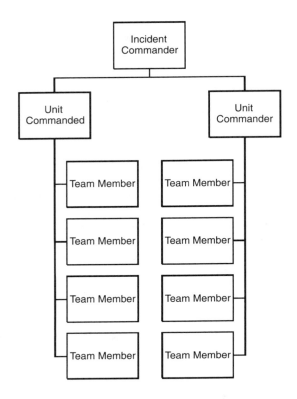

FIGURE 9.22 Incident command system with units.

INDUSTRY/FIRE DEPARTMENT COOPERATION

Cooperation between the fire brigade or emergency team and the fire department is essential to effective emergency control. This relationship must be developed prior to an emergency. The better the relationship that internal response personnel have with external responders, the better the overall emergency response will be.

During an incident, the fire department, once called, is usually officially in charge. Upon arrival, they will establish command of the incident. All internal response groups should then report through the fire department command structure.

Internal response groups should appoint an individual to act as liaison to the fire department and any other external response organizations. This person must be able to provide information concerning the incident. The fire department will want to know the specific location of the fire, the best access to that location, the nature of the emergency, and any other specifics that are available regarding the incident. Information should be provided as to what actions have already been taken by internal response personnel. The fire department personnel will also need to be informed of resources that are available. These may include personnel, installed fire protection systems, equipment, extinguishing agents, or anything else that may prove useful.

10 Coping with Fire

CHAPTER OBJECTIVES

You will be able to identify and explain:

- Issues regarding incident concerns
- Recovery process
- Incident documentation
- Public relations during an incident
- Incident evaluation techniques

You will also be able to:

- Prepare incident plans
- Take essential immediate actions
- Document an incident
- Deal with the news media
- Evaluate incident cause

WHY WHAT YOU DO IS IMPORTANT

What is done during an emergency can have a major impact on the consequences of that emergency. We have already discussed in some detail the emergency response aspects of dealing with the crisis. This chapter focuses on coping with the other issues that need to be handled during and immediately after an emergency to minimize the impact of that emergency.

The losses incurred from an emergency can be greatly magnified or significantly reduced based on the actions taken. By understanding the need for prompt, effective action and by ensuring that the necessary actions are taken, the impact of emergencies is minimized.

NOTIFICATIONS

WHO SHOULD BE NOTIFIED

Communications during an emergency are of critical importance. Many individual groups have a stake in the operation of your facility. These groups should be kept informed of the status of the operation. Any time a major emergency occurs, the impact

223

on the organization experiencing the emergency has a ripple effect on other associated organizations.

NOTIFICATIONS THAT MAY NEED TO BE MADE

The notifications that will be required depend on the size and scope of the emergency. If we assume a significant emergency has occurred that will have substantial negative consequences on the operation, many more groups will need to be notified than if the emergency was small and relatively insignificant.

Internal notification should always be made first. This would include employees, the corporate office if the facility involved is separate from the main office, the parent company if the facility is a subsidiary, unions, legal department personnel, and stockholders if the organization is a publicly held company. These people must be promptly notified of problems that may affect their situation.

Customers should also be notified early, particularly if delivery dates, product availability, or other customer-related items will be substantially affected. The sales force will generally be best prepared to notify customers. Suppliers and subcontractors should also be notified. Disruptions in your operations may have a major impact on those individuals.

Insurance representatives need to be notified promptly to allow them to prepare for your claim. The activities of the insurance company may require some time, so it is in your best interest to involve them early. Bankers and other financial institutions that have a relationship with your organization should be kept informed on the situation. They will find out eventually if financial issues are going to be a problem, and it is far better for them to hear it from you than from a secondary source.

The news media, and through them the general public, may also need to be informed. Dealing with the media is discussed in detail later in this chapter.

WHEN TO MAKE NOTIFICATIONS

Sooner is better than later as far as notifications of bad news are concerned. Delaying notifications creates the impression that no notification was planned. This can erode trust with the group involved and magnify the problem.

HOW TO HANDLE NOTIFICATIONS

The specific content of notifications should be based on the prospective audience. The notification for employees will not be identical to the notice given to customers. Notifications should emphasize the aspects of the emergency that affect the group being addressed. This is not intended to imply that information should be withheld from any group, but not all groups are interested in all information. Notifications should be designed to answer the questions that the individual group would be likely to ask. Employees will want to know if jobs will be affected, while bankers will be interested in the impact on financial issues like loan repayments.

PRIORITIZE SALVAGE EFFORTS

Even while an emergency is still in progress, the recovery efforts can be initiated. A recovery plan focuses on items relative to the emergency and is designed to get the operation back to normal as quickly as possible. Salvage and recovery efforts should be planned and prioritized to ensure that the most important activities receive attention first.

Many things may need to occur to recover from the emergency, and some of these activities may be completed simultaneously. Priority must be established for those items that have the greatest impact on returning to normal operations. For example, a machine that has been drenched by a sprinkler system must be cleaned promptly because it may become impossible to clean it properly later.

DOCUMENTING THE INCIDENT

Any incident at the facility should be thoroughly documented. If documentation is not done immediately after an incident, many details that may become important later will be forgotten and the opportunity to collect evidence will pass. Incidents should be documented to aid in determining the cause of the incident, creating preventive measures that could be instituted to prevent future incidents of this type, and providing a defense in the event of litigation involving the incident.

One of the first crucial things that should be done as part of the documentation phase is to interview all involved parties. The individual who discovered the emergency, those who responded to the emergency, and anyone else who operated at the scene of the emergency or who may have useful information should be interviewed as soon as possible after the emergency occurs.

The recollection that individuals have of the operation will begin to fade rapidly after the emergency. If these interviews are not conducted immediately, the information collected will be less useful than it might otherwise have been. Interviews should be conducted individually rather than in a group setting. Human nature will dictate that the strongly voiced recollection of a single individual may influence the recollection of others even if the vocal person is incorrect. Avoid making comments about what others have said during the interviews as this may also influence recollection and perception.

These interviews should be open-ended and focus on obtaining as much information as possible. There will be ample opportunity later to sort out the unimportant facts. Kipling's six honest serving men are a good basis for interviewing, and those are: who, what, where, when, why, and how. Avoid questions that may be answered with "yes" or "no." Do not ask leading questions during the interview process. For example, avoid asking "Was there thick black smoke?" A better question is, "What color was the smoke?" Suspend your own judgment about issues until the interview process is completed.

The entire incident scene should also be documented with photographs. Of particular importance are overall views of the entire area and close-up detailed views of extensive damage areas for possible sources of ignition. If the area of origin of the fire already been identified, extensive photographic documentation should be done in this area. Take as many photos as you think will be needed, then take more. It is easy to dispose of photos that are not needed, but it is difficult to deal with needing a photo that is not available. Once cleanup of the scene has begun, it is impossible to collect useful evidence later.

Incident reports and emergency response reports should be prepared as soon after the incident as possible. These reports present the information that has been collected in a useful review format. Even after these reports are completed, the original documentation should be kept on file. Obtain a copy of any reports prepared on the incident by others such as the fire department or insurance carrier.

DEALING WITH THE MEDIA

Any time your organization has a significant incident, the media will become involved. One of the most effective ways for dealing with the media during an emergency is to build a relationship with the people from the media prior to an emergency. By building a relationship of trust and cooperation with your local news media on an ongoing basis, your ability to work effectively with them during an emergency will be greatly improved.

The news media have an important and necessary job to do, with or without your help. The first important rule in dealing with the media is that you should give them the story. If the facts about the emergency are provided, second-hand information will not be needed. During major news events, a press conference or briefing should be held regularly, possibly as frequently as every 15 minutes during fast-paced periods. The timing of briefings should also take into consideration deadline constraints of the news media. For example, a press conference at 4:00 pm will be acceptable for most evening television newscasts but is past deadline for most evening newspapers. The newspaper deadline is usually 2:00 pm.

Another point in dealing with the media that cannot be overemphasized is the importance of honesty. Nothing will destroy your reputation with the media quicker than a lie. The media will almost always discover the truth eventually, which makes false information all the more damaging. Media personnel and the public will usually be considerably more understanding if a truthful account is provided from the beginning, even if some of the truth reflects poorly on the organization.

Another critical issue is off-the-record comments. Generally, off-the-record does not really exist. If the information provided is sufficiently important, it will show up on-the-record. Along these same lines, all cameras, tape recorders, microphones, etc. should always be considered on and operating.

If you are to be the primary spokesperson for the organization, it is an excellent idea to practice dealing with the media with someone from within your organization. Let them play the role of reporter to give you a feel for the circumstances you will be in during a media interview. If a video camera is available, videotape yourself during this practice.

HUMAN ISSUES

If the emergency has involved a significant injury or death, the human issues associated with the tragedy must be handled effectively. These incidents can be devastating to all personnel within the organization. Notification of families of workers that have been killed or severely injured is probably one of the most difficult tasks a manager will ever have to do. These notifications should be made as soon as possible and ideally handled in person. The telephone should be used only if it is the sole practical alternative such as in cases where the family member is distant from the facility location. Counselors should be available for family members and personnel at your facility.

Co-workers of the injured or killed employee will be particularly affected by the event. Another critical group is those that helped to deal with the emergency such as members of a first aid team. Recovery from losses of this type can be a long-term issue. Patience and understanding is required from all parties involved.

INSURANCE ISSUES

Insurance companies provide a valuable service that is needed by businesses. To get the most from this service it is important to understand the way insurance works. Insurance companies can be allies or adversaries after an emergency, and the activities within your organization can have a major effect on which they become. The insurance business is based on collecting a relatively small amount of money from many organizations and paying out as little of the money as possible in claims. When you purchase insurance you are gambling that you will experience a loss covered by that insurance policy. The insurance company uses the law of averages, based on studying loss-experience, to set the price of insurance. For example, if the insurance company's clients have loss-experience of $100,000 per year, the company charges premiums that will allow it to cover those losses, their operating expenses, and a reasonable profit. The insurance company does not have to predict who will have the loss, only that the losses indicated by probability are covered by the premiums they collect.

To keep premiums low and profits high, the insurance company wishes to minimize losses. This is also in the best interest of the insured.

When an emergency has occurred, the value of insurance is obvious, but getting maximum benefit from the insurance requires working with the insurance company from the beginning. Insurance agents should be consulted promptly and should be involved in the recovery process.

CAUSE DETERMINATION

The determination of the cause of an emergency is important so that lessons can be learned from the emergency. If the fire was arson, cause determination is the first step in the criminal investigation. Generally, fire cause determination should be left to experts in this field. The local fire department, the county or state fire marshal's office, and the insurance company should have personnel that can assist in this essential activity.

EVALUATE BUSINESS IMPACT

When a major emergency occurs, a thorough evaluation of the status of the business and the specific impact of the emergency is needed. This evaluation should identify the options that exist for recovery and assess the impact on short-, mid-, and long-term operations. An unbiased evaluation immediately after the emergency will improve the chances for a successful recovery. One of the dangers of not performing an evaluation is that significant items will be overlooked. For example, if the organization has a ninety-day note coming due in thirty days and that money is needed for the recovery expenses, negotiations with the bank for an extension should begin immediately. If the bank is not informed of the problem until the note is due, the organization will be much less likely to get an extension.

This business evaluation must look at the overall picture, not just the specifics of physical emergency. For example, how will the organization's customers be served? If the operation will be shut down for an extended period, can alternate manufacturing facilities or a subcontractor be used to supply customers in the interim? A customer, once lost, may be hard to recover.

Business interruption is one of the most costly portions of a major emergency. These issues may even be a problem with relatively small emergencies if the fire has shut down a critical part of the facilities operations.

Appendix A

Annotated Bibliography

1. Cote, A.E., Ed., *Fire Protection Handbook*, 18th ed., National Fire Protection Association, 1997.
   This is the most widely accepted reference book covering general fire protection. It is an excellent source for technical information on all aspects of fire protection. This publication is also available on CD-ROM.

2. Cote, A.E., Ed., *Industrial Fire Hazards Handbook*, 3rd ed., National Fire Protection Association, 1990
   This is an excellent reference manual for industrial fire protection. It contains sections on hazards common to most industries. There are also sections dealing with issues in specific industries.

3. *IRInformation*, Industrial Risk Insurers, 1991
   This two-volume set in three ring binder format provides an excellent reference source. The manual is divided into sections for easy use. An update service is available on an annual subscription basis.

4. DiNenno, P.J. Ed., *The SFPE Handbook of Fire Protection Engineering*, 2nd ed., National Fire Protection Association, 199
   This reference provides detailed information on the engineering aspects of fire loss control.

5. Colonna, P.E., Ed., *Introduction to Employee and Life Safety*, National Fire Protection Association, 2001.
   This is an excellent introductory reference for fire protection, life safety, and compliance issues.

CHAPTER 2 FIRE BEHAVIOR

6. Friedman, R., *Principles of Fire Protection Chemistry and Physics*, National Fire Protection Association, 1998,

CHAPTER 4 LIFE SAFETY

7. Lathrop, J.K., Ed., *Life Safety Code Handbook*, National Fire Protection Association, 2000.
    This is the most widely accepted code for life safety in buildings. Federal, state, and local regulations usually adopt sections of or this entire document as reference. This handbook contains the entire text of the standard and explanatory notes. This publication is also available on CD-ROM.

CHAPTER 6 INSTALLED FIRE PROTECTION

8. Bouchard, J.K. Ed., *Automatic Sprinkler Systems Handbook*, National Fire Protection Association, 1999.
    This is the most widely accepted code for automatic sprinkler systems. Federal, state, and local regulations usually adopt sections of or this entire document as reference. This handbook contains the entire text of the standard and explanatory notes.

9. Bryan, J.L., *Automatic Sprinkler and Standpipe Systems*, 3rd ed., National Fire Protection Association, 1990.
10. Bryan, J.L., *Fire Suppression and Detection Systems,* Glencoe Press, 1974.
    These two books by Bryan offer detailed information on systems in general. All major types are covered.

Another source of excellent information on installed systems is the manuals available from systems manufacturers.

There are many of other publications on many specific aspects of fire protection including numerous codes and standards. Several of the associations listed in Appendix B offer publications that may prove useful. In addition, a search may be conducted on the Books in Print Web site, http://www.booksinprint.com/bip/.

Appendix B

Resource List

ORGANIZATIONS

Many organizations are involved in fire protection and safety. This list includes addresses and phone numbers for a few of the better known groups. In addition to the organizations listed, you can usually find a local safety organization in your area.

Alliance of American Insurers
Phone: 630-724-2138 Fax: 630-724-2190
3025 Highland Parkway
Suite 800
Downers Grove, IL 60515
http://www.allianceai.org

American Burn Association
Phone: 800-548-2876 Fax: 312-642-9130
625 North Michigan Avenue
Suite 1530
Chicago, IL 60611
E-mail: info@ameriburn.org
http://www.ameriburn.org

American Industrial Hygiene Association
Phone: 703-849-8888 Fax: 703-207-3561
2700 Prosperity Avenue
Suite 250
Fairfax, VA 22031
http://www.aiha.org

American Insurance Association
Phone: 202-828-7100 Fax: 202-293-1219
1130 Connecticut Avenue NW
Washington, DC 20036
http://www.aiadc.org

American National Standards Institute
Phone: 212-642-8908 Fax: 212-398-0023
11 West 42nd Street
New York, NY 10036
http://www.ansi.org

American Petroleum Institute
Phone: 202-682-8000 Fax: 202-962-4776
1220 L Street NW
Washington, DC 20005
http://www.api.org/

American Risk and Insurance Association
Phone: 610-640-1997 Fax: 610-725-1007
716 Providence Road
P.O. Box 3028
Malvern, PA 19355-0728
http://www.aria.org

American Society for Industrial Security
Phone: 703-519-6200 Fax: 703-519-6299
1625 Prince Street
Alexandria, VA 22314
http://www.asisonline.org

American Society for Testing and Materials
Phone: 610-832-9585 Fax: 610-832-9555
100 Barr Harbor Drive
West Conshohocken, PA 19428-2959
http://www.astm.org

American Society of Safety Engineers
Phone: 847-699-2929 Fax: 847-296-9221
1800 East Oakton Street
Des Plaines, IL 60018-2187
http://www.asse.org

Automatic Fire Alarm Association (AFAA), Inc.
Phone: 407-322-6288 Fax: 407-322-7488
P.O. Box 951807
Lake Mary, FL 32795-1807
http://www.afaa.org/

British Safety Council
Phone: [44] (1817)-411231 Fax: [44] (1817)-414671
National Safety Centre
70 Chancellors Road
Hammersmith, London W6 9RS
United Kingdom
http://www.britishsafetycouncil.co.uk

Building Officials and Code Administrators Intl.
Phone: 708-799-2300 Fax: 708-799-4981
4051 West Flossmoor Road
Country Club Hills, IL 60478-5795
http://www.bocai.org/

Canada Safety Council
Phone: 613-739-1535 Fax: 613-739-1566
1020 Thomas Spratt Place
Ottawa, ON K1G 5L5
Canada
http://www.safety-council.org/

Canadian Association of Fire Chiefs
Phone: 613-270-9138 Fax: 613-599-7027
P.O. Box 1227
Station B
Ottawa, ON K1P 5R3
Canada
http://www.cafc.ca/

Canadian Automatic Sprinkler Association
Tel: 905-477-2270 Fax: 905-477-3611
335 Renfrew Drive
Suite 302 Markham, ON L3R 9S9
Canada
http://www.casa-firesprinkler.org/

Canadian Center for Occupational Health and Safety
Phone: 905-572-2981 Fax: 905-572-4419
250 Main Street
Hamilton, ON L8N 1H6
Canada
http://www.ccohs.ca

Canadian Centre for Emergency Preparedness
Phone: 905-546-3911 Fax: 905-546-2340
P.O. Box 2911
Hamilton, ON L8N 3R5
Canada
http://www.ccep.ca/

Canadian Fire Alarm Association
Phone: 905-944-0030 Fax: 905-479-3639
P.O. Box 262
Markham, Ontario L3P 2J7
Canada
http://www.cfaa.ca/

Canadian Fire Safety Association
Phone: 416-492-9417 Fax: 416-491-1670
2175 Sheppard Avenue East
Suite 110
Willowdale, ON M2J 1W8
Canada
http://www.canadianfiresafety.com/

Canadian Society of Safety Engineers
Phone: 905-893-1689 Fax: 905-893-2392
P.O. Box 294
10435 Islington Avenue
Kleinburg, ON L0J 1C0
Canada
http://www.csse.org

Central Station Alarm Association
Phone: 703-242-4670 Fax: 703-242-4675
400 Maple Avenue
Suite 201
Vienna, VA 22180
http://64.226.207.204/index.html

Certified Fire Protection Specialist Board
Phone: 617-984-7484 Fax: 617-984-7056
1 Batterymarch Park
P.O. Box 9101
Quincy, MA 02269
http://cfps.nfpa.org

Compressed Gas Association
Phone: 703-788-2700 Fax: 703-934-1831
4221 Walney Road,
5th floor
Chantilly, VA 20151-2923
http://www.cganet.com

Fire Protection Association
Phone: [44] (020) 790-25300 Fax: [44] (020) 790-25301
Bastille Court
Paris Garden, London SE1 8ND
United Kingdom
http://www.thefpa.co.uk

Fire Protection Association Australia
Phone: [61] 03-9890-1544 Fax: [61] 03-9890-1577
13 Ellingworth Parade
P.O. Box 1049
Box Hill, Victoria 3128
Australia
http://www.fpaa.com.au

Industrial Accident Prevention Association
Phone: 416-506-8888 Fax: 416-506-8880
250 Yonge Street
28th Floor
Toronto, ON M5B 2N4
Canada
http://www.iapa.on.ca/

Institute of Fire Engineers
Phone: [44] (5335)-53654 Fax: [44] (5334)-71231
148 New Walk
Leicester, LE1 7QB
United Kingdom
http://www.ife.org.uk

Institute of Occupational Safety and Health
Phone: [44] 01162573100 Fax: [44] 01162573101
The Grange
Highfield Drive
Wigston, Leicestershire LE18 1NN
United Kingdom
http://www.iosh.co.uk

Insurance Information Institute
Phone: 212-669-9200
110 William Street
New York, NY 10038
http://www.iii.org

International Association of Fire Chiefs
Phone: 703-273-0911 Fax: 703-273-9363
4025 Fair Ridge Drive
Fairfax, VA 22033-2868
http://www.iafc.org

International Association of Practicing Safety Engineers
P.O. Box 151
Baldwinsville, NY 13027-0151
http://www.iapse.org

International Safety Equipment Association
Phone: 703-525-1695 Fax: 703-528-2148
1901 North Moore Street
Suite 808
Arlington, VA 22209
http://www.safetycentral.org/isea

International Society of Fire Service Instructors
Phone: 800-435-0005 Fax: 540-899-2178
P.O. Box 2320
Stafford, VA 22555
http://www.isfsi.org

National Association of Fire Equipment Distributors
Phone: 312-245-9300 Fax: 312-245-9301
401 North Wabash Avenue
Suite 732
Chicago, IL 60611
http://www.nafed.org

National Association of Fire Investigators
Phone: 941-359-2800 Fax: 941-351-5849
7678 15th Street East
Sarasota, FL 34243
http://www.nafi.org

National Fire Protection Association
Phone: 617-770-3000
One Batterymarch Park
P.O. Box 9101
Quincy, MA 02269
http://www.nfpa.org

National Safety Management Society
Phone: 800-321-2910
http://www.nsms.ws/

National Safety Council
Phone: 630-285-1121 Fax: 630-285-1315
1121 Spring Lake Drive
Itasca, IL 60143
http://www.nsc.org

National Safety Council of Australia
Phone: [61] 02-9666-4899 Fax: [61] 02-9666-4811
P.O. Box 810
Mascot, NSW 2020
Australia
http://www.nsca.org.au

Royal Society for the Prevention of Accidents
Phone: [44] (2170)-68121 Fax: [44] (2176)-54295
22 Summer Road
Acocks Green, Birmingham B27 7UT
United Kingdom
http://www.rospa.co.uk/cms/

Safety Institute of Australia
Phone: [61] 03-9890-6304 Fax: [61] 03-9890-6310
P.O. Box 93
Box Hill, VIC 3128
Australia
http://www.sia.org.au

Society of Fire Protection Engineers
Phone: 301-718-2910 Fax: 301-718-2242
7315 Wisconsin Avenue
Suite 1225W
Bethesda, MD 20814
http://www.sfpe.org

System Safety Society
Phone: 703-450-0310
P.O. Box 70
Unionville, VA 22567-0070
http://www.system-safety.org/

U.S. GOVERNMENT AGENCIES

Chemical Safety and Hazard Investigation Board:
   http://www.chemsafety.gov/
Consumer Product Safety Commission: http://www.cpsc.gov/
Federal Emergency Management Agency: http://www.fema.gov
Fire Administration: http://www.usfa.fema.gov/
Mine Safety and Health Administration: http://www.msha.gov
National Institute for Occupational Safety and Health:
   http://www.cdc.gov/niosh/homepage.html
Occupational Safety and Health Administration: http://www.osha.gov

Individual U. S. states also typically have Web sites that may be used to obtain information about regulatory requirements within a particular state.

SELECTED WEB SITES

There is an overabundance of useful information on the Internet. The Web sites listed below are a small sampling of what is available.

Amerex Corporation: http://www.amerex-fire.com
American Water Works Association: http://www.awwa.org
Ansul Incorporated: http://www.ansul.com
Badger Fire Protection, Inc.: http://www.badgerfire.com
British Standards: http://www.bsi-global.com/group.xalter

Canadian Standards Association:
   http://www.csa-international.org/default.asp?language=english
Central Sprinkler: http://www.tyco-central.com/
Eagle Manufacturing Company: http://www.eagle-mfg.com
European Agency for Safety and Health at Work: http://europe.osha.eu.int/
Fenwal Protection Systems: http://www.fenwalfire.com
Fike Corporation: http://www.fike.com
FIRECON: http://www.FIRECON.com
FM-200: http://www.fm-200.com/
FM Global: http://www.fmglobal.com/default.asp
GEM Sprinkler: http://www.tyco-gem.com/
General Fire Extinguishers Corp.: http://www.genfire.com
Grinnell: http://www.tyco-gem.com/
Halotron: http://www.halotron-inc.com
Industrial Risk Insurers: http://www.industrialrisk.com/
Japan Industrial Safety and Health Association:
   http://wellmet2.wellmet.or.jp/~jisha/indexe.htm
Justrite Manufacturing Company: http://www.justritemfg.com/
Kidde: http://www.kidde-fire.com/
Reliable Automatic Sprinkler Company: http://www.reliablesprinkler.com

Index

C

M